REINVENTING LIFE

A Guide to Our
Evolutionary Future

Jeffrey Scott Coker

THE BLUE HELIX
Charleston, SC

For my parents and children,
and for yours,
in hopes that we will learn
from our evolutionary past
to create a better future.

Contents

Part I

Introduction

1

FROM JAGUARS TO AVATARS

How Are We "Reinventing Life"?

It is as if man had been suddenly appointed
managing director of the biggest business of all,
the business of evolution—appointed without
being asked if he wanted it, and without proper
warning and preparation. What is more, he
can't refuse the job. Whether he wants to or not,
whether he is conscious of what he is doing or
not, he *is* in point of fact determining the future
direction of evolution on this earth. That is his
inescapable destiny, and the sooner he realizes it
and starts believing in it, the better for all
concerned.

—Julian Huxley (*New Bottles for New Wine*, 1957)

I was deep in the Amazon rainforest of Peru, spending
the day with four of my students who were doing
experiments on leaf-cutter ants. Because they didn't
need my help at that moment, I walked down the trail

about forty yards to look around. All of a sudden, a troop of monkeys screamed frantically in the distance and then a nearby flock of birds called wildly. Something was coming. I heard a crunch in the undergrowth. Then, to my surprise, the unmistakable deep growl of a big cat rumbled through the forest. It was close.

I hurried back to warn the students. As soon as I returned, there was another crunch in the bushes. This time it was close to our group. "There is something big out there," I said. We all paused for a moment and then heard the same growl from the thick undergrowth nearby. "Oh my God, that sounds just like a big cat," said one girl. It had clearly followed me, and quickly.

The two male students, not realizing the seriousness of our situation, took several steps toward the growl to get a better look. A more intense growl and my plea to stay back quickly changed their minds. Then there was silence. We stood still and waited, staring into the dense foliage, knowing that running would make us look like prey. Until you're in a situation like this, it's hard to comprehend the simultaneous thrill and horror of being stalked by a more powerful animal. In the jungle, a big cat has nearly every advantage—speed, stealth, and natural weaponry. Our one advantage was being in a group of five.

After a long wait, everyone felt that the cat had moved on. As we picked up our gear and prepared to leave, a student stomped his foot near an entry to an ant colony. They had been doing experiments on ant behavior, and the desire to finish the day's work proved irresistible. The stomping prompted an angry, earth-shaking growl, this time louder and longer than the others. The cat hadn't gone anywhere. So much for the conventional wisdom that big jungle cats are reclusive and secretive. This one was defending its territory, and we were in the wrong spot. We quickly walked back down the trail to our camp.

The next day, as we traveled down the Madre de Dios River by boat, we saw a jaguar and two cubs peeking at us from the brush.

Those few moments in the Amazon seemed to wake neurons in my brain that had never been disturbed — ancestral memories of pure wilderness and a time when humans were not atop the food chain. Not so long ago, the world was in a wild and natural state, the product of a three-billion-year evolutionary process. It was both spectacular and harsh on a scale that modern humans can hardly imagine. This is a world we've mostly left behind.

Today, there are few natural refuges left that even vaguely resemble their past splendor. My students and I had to travel halfway around the world to find such a place. But even in the Amazon, the older locals remember childhoods when the animals were bigger, more abundant, and less skittish. Even in the few remaining refuges, humans have already had profound impacts.

Most people in the modern world have never experienced the thrill and horror of encountering a big cat in its natural habitat. Instead, we have docile lap cats, domesticated animals that we breed to suit our fancy. Instead of wilderness, complete with large animals like jaguars, we have the cities, suburbia, and farmland of human development. Even the fields, forests, and oceans we think of as "natural" are usually utterly transformed versions of what used to be nature. We have traded a world shaped by unconscious evolutionary forces for one driven largely by human behaviors, imaginations, and fallacies. The future of life is now governed by decisions — our decisions.

The power and dominance of the jaguar were very real to us that day in the Amazon, just as they were real to our ancestors. But from a modern global perspective, the jaguar is at our mercy. It will live or die based on decisions we make.

What should we do with jaguars? Should we conserve their forests and let them thrive? Should we hunt and kill them for economic gain? Should we keep them in captivity? Should we domesticate some of them, like we have other cats, perhaps using genetic engineering to make them smaller (and less threatening)? Should we allow the whims of markets and economies to decide their future? If they go extinct, should we use cloning to resurrect them? Should we use their stem cells to grow fur for clothing? Should we use their genes to enhance our own speed, strength, or disease resistance? Should we create jaguar-like avatars using computers so that they can become digitally immortal, perhaps joining humans in the evolutionary leap from biological to digital consciousness?

Whatever the future of jaguars, we decide it. Whatever the future of life, in general, we decide. This profound responsibility is my reason for writing this book, and should be your reason for reading it.

A Culture of Obsession and Denial

It might be tempting to think about the reinvention of life as someone else's doing. Perhaps you would like to put all the responsibility on fossil fuel industries that are contributing to climate change, scientists who are creating new technologies, or politicians who are governing societies. These groups are responsible, but so are you. Our modern culture is built around the management and manipulation of Earth's biology. In fact, we do our best to assert control over the biological world every day.

We each spend inordinate amounts of time bathing, cleaning, and grooming to control the microorganisms all around us. In our landscapes, we insist on selecting the placement of nearly every plant, giving the plants that grow on their own the derogatory name "weeds." Since we all contribute to the incredible amounts of fossil fuels burned every day, we are all responsible for changing the Earth's climate, which in turn is reshaping ecosystems worldwide. We have created new environments — roads, lawns, building interiors, reservoirs, and farms — that drive waves of evolutionary events. We use medical technology to greatly extend our life spans. We treat cats and dogs like people and make it illegal to harm them, while we simultaneously imprison and eat pigs and cows. We spray chemicals to kill some organisms, and other chemicals to help other organisms grow. Nearly every consumer choice we make has an evolutionary impact — our stores are filled with products made or derived from living things, from cotton clothing to leather shoes. We even build and purchase machines to give us powers that transcend our biological condition, from communicating across vast distances to traveling 70 miles per hour. It is not just a few people who are reinventing life — all of us are.

The reinvention of life also dominates our everyday consciousness. Consider the toys we buy for children. From birth, we surround them with stuffed animals. Some represent human destruction and the species humans have endangered — jaguars, bears, tigers, penguins, apes, seals, and other large animals. The rest represent human creation, including many stuffed toys that aren't real animals at all — dragons, unicorns, life-like robots, and anime figures. Why do we make our children's rooms look like a cross between a hunter's trophy room and a mad scientist's laboratory? Most parents probably intend a message of connection with nature, but the message is wrapped with a bow of control. It's as if the very first lesson we want to teach future generations is that humans can both destroy and create life. As children outgrow stuffed animals, the message becomes more blatant. Walk through a department store toy section and you'll notice that nearly every product represents an enhanced biological entity — superheroes, cyborgs, talking animals, human–animal hybrids, and intelligent cars.

Likewise, some of our most popular sports are already dominated by performance–enhancing drugs and other types of enhancements. Anyone who has watched an NFL football game on the Fox channel has seen a robot-like football player avatar during commercial breaks. He stretches, runs in place, and jumps around like a rookie on the sidelines just waiting for his turn to get into the game. He seems to be begging for acceptance, "Just imagine what the game would be like if I could play!" Is that what future professional athletes might look like? What are we preparing ourselves for?

More evidence comes from Hollywood. Movies are often a subconscious way of preparing ourselves for radical futures that may lie ahead. Movies are like a testing ground for our evolutionary options.

Our favorite films include people with enhanced abilities or superpowers, animals that talk, environments that have been completely transformed, or some other radical biological transformation. In fact, over 90 percent of the 50 highest grossing films of all time involve characters with qualities that far exceed their typical biological situation. We are fascinated by exotic environments and the ability to become another species (Avatar), children with magical powers (Harry Potter), bringing dinosaurs back to life (Jurassic Park), using "the force" to alter the environment with our mind (Star Wars), superheroes (Spider-Man, Iron Man, Batman, and others), and trans-species communication (most Disney movies). With only a few exceptions, stories taking place within the confines of everyday human existence are less popular than biological fantasies. (The ironic exception is one of the highest grossing films of all time, Titanic, which is the ultimate parable about how human technological excess can lead to total disaster.)

Clearly we have a cultural obsession with the reinvention of life. It dominates our hygiene, eating, shopping, entertainment, and nearly everything else.

And now for the hypocrisy. . . At the same time that we dominate the evolutionary process and struggle for ever-greater control over nature, we have also created cultural myths that deny our evolutionary impacts. Culture provides several very appealing, yet deeply flawed ideologies that allow people to reject accountability for their actions.

Most obvious is anti-evolutionism. In a world where humans dominate evolution and shape the future of life, many don't even believe that evolution happens. As we explore throughout this book, evolution underlies many of our biggest global challenges, from hunger to environmental change. For example, more than 20 percent of human deaths result from evolving microbes. Anti-evolutionism directly contributes to disease, poverty, and environmental destruction by making much of our collective brainpower inaccessible for solving global problems. (We'll discuss evolution more directly in chapter 2.)

Others embrace fairy-tale notions of chance and fate. "We shouldn't mess with life, because fate will decide." The lovers of fate prefer to think that how the future turns out has little to do with our decision-making or work ethic. For example, consider the notion that everyone is born with "one true love," and after we find our predetermined person we'll live happily ever after. This is like a Santa Claus story for teenagers and adults. Love and marriage are not fated, and it turns out that you have to work at them. Our decisions matter. We can and will change the future, whether through action or inaction. More specifically, our collective actions will determine the course of evolution and the future of life.

Other denialist ideologies abound. "We don't need to worry about global problems, because the market will fix it." "There's no need to worry about how we behave, since there's no way we could stop X from happening anyway." "We don't have to worry about a thing, since future technologies will solve everything." "Environmental destruction isn't a problem because the end of times is near anyway." Let's be honest with ourselves. These are just excuses for apathy and inaction. We can do better.

Reality does not change based on what you believe. The reality of the modern world is that evolution is happening, humans are driving it, and the choices that we make today will shape the future of life.

Embracing our Evolutionary Future

This book shows myriad ways that we are changing life, explores our options for the future, and then embraces a broader philosophy—we must accept our responsibility as stewards of evolution. We should embrace evolutionary stewardship and take care of life with compassion and thoughtfulness. When we think about evolution and make decisions with long-term consequences in mind, we can save millions of lives and trillions of dollars each year. We can make drugs last longer, cure deadly diseases, greatly reduce hunger, preserve and restore wilderness, expand human abilities, and ultimately spread life to other worlds.

Some well-intentioned people will immediately protest that we have no right to direct life's future— "We shouldn't play God." In reality, it doesn't matter whether we want to direct the future of life—we have no choice. Billions of humans on a single planet cannot help but direct life's future. In fact, each of us makes decisions every day that are essentially votes for one evolutionary outcome or another. The real issue, then, is whether we think about what we're doing. Which is better, randomness or intentionality? When we are oblivious to evolutionary consequences, the results are almost always destructive for humans and other species. The outcomes of not consciously directing evolutionary processes include persistent disease, poverty, a degraded environment, and the stifling of human progress. This is an awfully high price to pay for ideologies of denial and randomness. In terms of practical and measurable outcomes, evolutionary stewardship is the most ethical and responsible path forward.

Scientists have only a small role in determining the future of life. That is why this book is written for all thoughtful people, those who care about how our actions impact the future. I would hope that this includes students, parents, CEOs, community leaders, journalists, educators, scientists, and many others. The future of life is being determined largely by our policies, ethics, consumer choices, reproductive decisions, and everyday behaviors. Thus, it is profoundly important that we consider the implications of our actions and, as Gandhi suggests, be the change we wish to see in the world.

In *The Dragons of Eden*, Carl Sagan summarizes the modern era by saying, "We live in a time when our world is changing at an unprecedented rate. While the changes are largely of our own making, they cannot be ignored. We must adjust and adapt and control, or we will perish." How can we be responsible stewards of evolution? If we planned a scientifically and ethically responsible evolutionary path for the future, where might we go?

2

EVOLUTION ALL AROUND US

Evidence for Our Evolutionary Past, Present, and Future

Al this mene I by love, that my felynge
Astonyeth with his wonderful werkynge
So sore y-wis, that whan I on him thinke,
Nat woot I wel wher that I flete or sinke.

—Geoffrey Chaucer

The epigraph by Chaucer is incomprehensible to most modern English speakers. And yet it is "English," or at least what English looked like in the fourteenth century. If you follow modern English back in time through Middle English, Late Latin, Early Latin, Greek, and beyond, it becomes obvious that language is constantly evolving. A modern English speaker could not hold an effective conversation with an English speaker from 600 years ago, nor with an English speaker 600 years in the future. A letter may change here and a sound there in small, nearly imperceptible ways. Over time, these small changes accumulate until one language evolves into another.

We see similar evolutionary processes at work throughout our lives. Airplanes evolved from rickety wooden gliders to the mammoth commercial airplanes and agile fighter jets of today. Similar changes occur with cars, telephones, televisions, fashion, medicine, building construction, and other aspects of society. On a cosmic level, stars, planets, and galaxies evolve in dramatic ways. Evolution is part of the fabric of everything. As far as we know, everything changes over time except natural laws, and even that is not certain.

For over 3 billion years, living things on Earth have steadily evolved in response to their environment. Although this book is about change in today's world, the evolutionary history of life provides the essential context. Evolution is not just a historical record. Instead, it is the fabric of life on which the modern story is woven. To understand where life is going in the future, we must first understand where it's already been.

The Big Picture of Evolution

In the context of life, "evolution" refers to descent with modification. In other words, traits within a species can change from generation to generation. On a small scale, "evolution" means that gene frequencies within a population change in response to environmental pressures. On a large- scale, "evolution" means that species descend from other species over many generations.

Evolution immediately suggests thinking of life as a tree. The base of the tree represents early unicellular life—the common ancestor for modern species. Moving up the tree represents moving through 3 billion years of history. New species come about as branches bifurcate into multiple branches. Some branches stop as species go extinct, whereas other branches proliferate. The outer branches represent modern species, which now enjoy their day in the sun. (Modern molecular biology suggests that the tree is also covered with vines that allow some genes to be shuttled across branches, but we'll save that complexity for later.)

Evolution in nature requires three core ingredients. First, there must be diversity. As we'll discover later, biological diversity is created by mutations in our DNA, our genetic code, that accumulate over time. Second, there is a selection process that eliminates some diversity and leaves other diversity. Selection might result from predation, camouflage, size, shape, eyesight, speed, physiology, or many other things. Third, the organisms carrying the remaining diversity reproduce and pass their genes to future generations. Thus, evolution progresses mostly through a cyclical process of creating diversity, eliminating some of it, passing along the rest, creating more diversity, eliminating some of it, and passing along the rest. In this way, life has adapted itself to environments all around the world.

When Charles Darwin and Alfred Wallace first proposed evolution through natural selection in the mid-1800s, their evidence was powerful but somewhat limited. Their early arguments were almost entirely based on observations that could be made in nature with the naked eye. Much knowledge that we now take for granted did not exist at that time, such as the nature of genes, the existence of most dinosaurs and other species of previous ages, and a multitude of other discoveries that are now fundamental to our understanding of nature.

In the modern world, there are hundreds of lines of evidence to support evolution. Millions of research articles are published every year that deal with evolution, each adding small bits of information to our understanding of the evolutionary tree.

Ten Lines of Evidence for Evolution

It would take libraries of writing to exhaustively list all of the evidence for evolution, and that is not my goal here. Instead, let's examine 10 lines of evidence that end up being especially useful for understanding modern biological change. These will serve as a starting point for understanding how we are driving evolution into the future.

First, we have selectively bred (evolved) food crops and many ornamental plants. The last 10,000 years of human history is a testament to how species can be modified through selection. Virtually everything you see in the produce section of a grocery store has been selectively bred. Some, like corn, have been modified steadily for thousands of years, whereas others are much more recent creations. A good example of the power of evolution is the breeding of a species of wild mustard (*Brassica oleracea*). Using naturally occurring mutations in this species, humans have bred broccoli, cauliflower, cabbage, kale, kohlrabi, and other vegetables. To our everyday sensibilities, these foods neither look nor taste the same. Yet they sit adjacent to one another on a small branch of the evolutionary tree.

Second, we have selectively bred (evolved) our pets and livestock. Virtually every animal that we commonly come into contact with has been selectively bred. As with crop plants, the animals we eat have all been extensively modified. Those on land (i.e., cows, pigs, and chickens) have been modified for thousands of years, whereas those in the ocean (i.e., fish and shellfish used in aquaculture) were bred mostly in the past 200 years. Most people best appreciate the evolutionary trees of pets, especially cats and dogs. From a small pool of ancestral dogs (wolves), the wide diversity of modern dog breeds was created. We have huge Irish wolfhounds and tiny chihuahuas, slender greyhounds and stumpy bulldogs, some friendly breeds and others known for their ferocity. Evolution has been used as a tool to produce friends, protectors, cuddlers, hunters, or whatever else people want.

Third, natural selection can be seen on an annual basis. Anyone from an agricultural background knows that you can't use the same pesticide indefinitely. Imagine spraying a field where 99 percent of an insect species is killed by a pesticide, but 1 percent have a DNA mutation that make them resistant. The pesticide will be very effective for a year or two. But every year, those resistant insects make up a bigger proportion of the population. They keep passing their genes to the next generation and, without as much competition, they proliferate. After a few years, much of the population is resistant and the pesticide is less effective. If a farmer keeps using the pesticide after it becomes less effective, it will quickly become useless as the whole population becomes resistant. The same is true for herbicides, fungicides, and vaccines. The reason that we have to get a flu shot every year isn't because the drugs wear off, but because the influenza virus evolves extremely quickly. Researchers in agriculture and medicine are in a constant evolutionary arms race.

Fourth, there are thousands of documented examples of natural selection in nature. Scenarios similar to the pesticide example play out in every natural ecosystem on Earth virtually all the time. Scientists have documented tens of thousands of examples of natural selection in action. Most people think of natural selection in predatory terms — slow zebras are eaten by lions, whereas faster zebras escape and reproduce, or noncamouflage moths are eaten by birds while camouflage moths escape. Most natural selection is far more subtle. It usually only takes a small environmental change to give some organisms an advantage over others. Climate, nutrient levels in the soil or water, the amount of carbon dioxide in the air, the presence of another species, or innumerable other factors can tip the scales of natural selection and have major evolutionary implications over time.

Fifth, an organism's body can be radically changed by a mutation in just one gene. Small-scale evolution within a species is easy for anyone to imagine. Large-scale evolutionary events that lead to one species becoming another species may seem more difficult. How can a species get another set of appendages or some complex structure? Wouldn't there have to be intermediate forms that are still functional? Sometimes there have been intermediate forms leading up to complex structures such as eyes. You can look through the animal kingdom and see much of the progression from simple photoreceptors to complex organs. However, many people are surprised to hear that huge evolutionary changes can happen very quickly. A single DNA mutation in homeotic genes, which determine how a body is formed, can result in a whole new limb or organ. For example, mutating one gene can produce a second set of wings in a fruit fly, can lead to legs growing where antennae would normally grow, or a multitude of other possibilities. Although we rarely see them, all sorts of similar mutations happen in humans — babies are born with hands and feet in the wrong place, tails, different numbers of fingers and toes, webbed feet, and so on. All it takes is one mutation to the right gene, setting off a cascade of genetic and biochemical pathways that ultimately cause a big structural change. Also, some genes regulate the mutation rate of other genes. Thus, a small number of mutations, even just one, can have far-reaching evolutionary implications.

Sixth, all species share characteristics. The idea of evolution suggests that all of life has a common ancestor. If this is the case, then there should be some similarities between all living things that result from this common ancestry. And indeed, there are many. All organisms are made up of cells, contain DNA, and share similar chemical and biochemical pathways. On an even more fundamental level, all organisms are carbon-based. Thus, all species on Earth share common fundamental characteristics. This has remarkable evolutionary implications in the modern world. For example, because all organisms pass down information using DNA, traits can be transferred from species to species by cutting and splicing DNA.

Seventh, body structures show evolutionary

relationships. To our everyday sensibilities, a human arm seems completely unlike a bat's wing or a sea lion's flipper. Look more closely, however, and you will find that the organization of bones in all three (and in other mammals) is very similar. Individual bones are different sizes and shapes, of course, but size and shape turn out to be easily modified through DNA mutations. Differences between species that appear on the surface to be huge often represent relatively small modifications to the body plan of a common ancestor.

Eighth, DNA sequences show evolutionary relationships. You may have heard before that the DNA sequence (the string of A's, C's, T's, and G's) of human genes is 99 percent the same as the DNA of chimpanzee genes, 98 percent the same as a gorilla's, and so forth. More closely related species have more similar DNA sequences, and more distantly related species have less similar DNA sequences. Thus, one can track evolutionary relationships by comparing DNA sequences.

Ninth, the fossil record shows that organisms change over time. Paleontologists and geologists have amassed a huge collection of fossils from different locations around the world, allowing us to see glimpses of Earth's past. We now know that most of the species that existed millions of years ago were very different from those of today. Digging through successive layers of sediment shows that species evolve into new species, and there are periods where biodiversity increases dramatically. Likewise, species go extinct; there have been at least five mass extinctions in the past. In fact, over 95 percent of species that once existed are no longer here.

Finally, evolution is consistent with the major

ideas of other scientific disciplines, namely, plate tectonics and the Big Bang theory. The story of the universe that science has uncovered is very cohesive. Astronomers have found that the Big Bang occurred around 13.7 billion years ago, galaxies began forming about 300,000 years after that, and then our sun and Earth formed around 4.6 billion years ago. By examining rocks, geologists confirm that the Earth is around 4.6 billion years old. They have also shown that the Earth has continental plates that move around, which has evolutionary consequences because it changes the physical environment. Meanwhile, paleontologists and biologists have found that life is around 3.8 billion years old, and they have uncovered much of the life that has evolved up to the present time. This remarkably coherent timescale adds context to how we think about modern life. Furthermore, it suggests much about the future—the universe will continue to expand, the Earth's plates will continue to shift around, and life will continue to evolve.

Evolution as a Worldview

Evolution has many important consequences for how we view modern life, including how we perceive ourselves. First, it suggests that all species are related. We are related to oak trees and leopards and fish (albeit with millions of years between our evolutionary branches). Depending on your mindset, you might find this a beautiful or detestable thought. Personally, I find it quite beautiful.

Evolution also suggests that all species are interconnected. Species did not evolve in isolation.

Instead, each co-evolved in ecosystems in which it was surrounded by other evolving species. As a result, modern changes to a species ripple through an ecosystem, often in unpredictable ways.

Another meaningful part of evolutionary history is that humans are a youthful species. We have been around far less than 1 percent of the time since life came about and have been on Earth for far less time than many other species. We've just barely gotten started.

In summary, evolution suggests that humans are a part of a highly interconnected and interdependent world. We are one with nature, and we share the fate of our environment.

Evolution suggests one more thing—and this really is the most important. As Charles Darwin notes in *Origin of Species*, "Judging from the past, we may safely infer that not one living species will transmit its unaltered likeness to a distant futurity." Although it may be tempting to view modern life as the final result of evolution, life continues to evolve. The tree of life is still growing.

The difference between evolution over the past 3 billion years and evolution in the modern world is that now the activities of a single species dominate the process. When humans first began to stride across African savannahs and then migrate to other regions, they found incredible biodiversity before them. It was a world full of large and wondrous plants and animals which had evolved intricate ecological relationships with each other over millennia. But once humans arrived, the reinvention of life soon began.

PART II
ECOLOGICAL CHANGE

3

OUR GREATEST IRONY

Human Influence and Dependence

Humanity i love you because you
are perpetually putting the secret of
life in your pants and forgetting
it's there and sitting down...

—e.e. cummings ("Humanity i love you," 1925)

*I*t would be ideal if we could coexist harmoniously with naturally evolved ecosystems. In reality, we must acknowledge that some degree of destruction and change is necessary for human civilizations to exist. Civilizations change environments and accelerate evolution by their very nature.

Even the simplest forms of civilization require some degree of human influence over nature. For towns and cities to emerge, food must be harvested and managed from fields, farms, fisheries, and other systems. Buildings require wood from forests or stone and metals dug from the earth. Energy for cooking, heating, and lighting often comes from materials that are mined or drilled. Clean water for drinking, irrigation, and sanitation is piped from lakes or rivers and then released back into the environment after use. Our day-to-day needs are intimately entwined with ecosystems all around us. We are totally dependent on natural resources, yet extracting them influences (and often threatens) their existence — this is the great irony of civilization.

Because we depend on the natural environment and influence it so heavily, it is obvious that we should try to achieve balance between what we take and what we put back. When societies lose balance, people suffer. When things get way out of balance, societies tend to fall apart.

It is not surprising that a large number of global problems related to poverty, politics, and violence are closely connected with environmental problems. Most North Americans are familiar with the plight of Haitian refugees who attempt dangerous ocean voyages to the United States in search of a better life. But what caused Haiti to become the poorest nation in the Western Hemisphere? To begin with, deforestation has ravaged the small island nation. Cutting down too many trees for charcoal has eroded soil from mountainsides, polluting waterways and making it difficult to grow crops. Likewise, overfishing has destroyed major fisheries on the Haitian coast. The resulting poverty has been severe, leading to general societal instability and violence. Obviously, Haiti's problems are not solely ecological. A history of dictatorship, government corruption, and poor education contribute to instability. Nevertheless, the nation's environment has been devastated to the point where the rest of society cannot function properly. Ecological stability and societal stability are intimately linked.

Other examples of the influence–dependence tension abound. Water disputes have caused unrest on every continent as countries dam up rivers and pull water for irrigation that would otherwise flow downstream into other countries. Bad irrigation practices throughout the world have deposited salt onto farmland, converting usable land into deserts. Fishermen off the coast of Somalia became pirates after international fishing vessels repeatedly overfished their waters, threatening their livelihood. Overpopulation in Bangladesh has caused desperate poverty. Illegal immigration into the United States is driven by combined ecological and economic stresses in Mexico and Central and South America. All of these examples support the fact that societal stability is impossible without healthy ecosystems to meet basic human needs. We must change natural ecosystems for civilization to exist, but we destroy them at our peril.

We Are the Environment

People often refer to "the environment" in distant terms, sometimes even asking us to choose between humans and the environment. This is completely illogical. Humans do not, and cannot, exist outside an environment. In fact, noting that we are "entwined" or "connected" to the environment still doesn't do justice to reality. In the most literal sense, we are the environment. Air, water, and soil—these are the three basic parts of our biosphere. Not only do we live in air, water, and soil, they also live in us. We breathe the air. We drink the water. And we eat food that has been grown in the soil. These things flow in and out of our bodies every day. Therefore, when we change parts of the biosphere, we change ourselves in the process.

As an example, consider the release of toxic materials from nuclear weapons. Many younger people know only about the bombs dropped on Hiroshima and Nagasaki during World War II. What is less well known is that, from 1945 to 1992, over 2,000 nuclear weapons were detonated as "tests." Over 1,100 of these were detonated by the United States on islands in the Pacific and Atlantic Oceans, as well as in Nevada, Utah, New Mexico, Alaska, Colorado, and Mississippi. We tested nuclear weapons above ground, underground, on the ocean's surface, under water, low in the atmosphere, high in the atmosphere, and even in space. Each time a bomb exploded, it obliterated a natural environment and spewed radioactive materials into the air, water, and/or soil. Everyone on Earth during the early days of nuclear testing carried radioactive fallout in their bodies in the forms of iodine-131, cesium-137, and several dozen other isotopes. For example, iodine-131 was carried by winds all over the world, landed on plants that were consumed by livestock, and was then consumed by humans in the form of milk and other products.

You can see the legacy of our decisions in our bodies and our societies. Environmental destruction is, in fact, self-destruction. Likewise, environmental restoration is self-restoration.

We benefit from dozens of "ecosystem services" —services that natural ecosystems provide to us for free. These include a wide range of essentials that we usually take for granted: air and water purification, soil formation, wood, fiber, crops, livestock, fisheries, fuel, medicines, nutrient cycling, pollination, climate regulation, and more. Ecosystems also provide incalculable benefits for the human spirit in terms of beauty, education, recreation, and spirituality.

In other words, natural ecosystems provide the fabric of our existence. When we cut down a forest, we gain wood and possibly usable land, but we lose the forest's ability to clean water, maintain wildlife, and pull carbon dioxide from the air. When we overfish a region of the ocean, we receive a short-term economic gain, but we reduce the ability of the fishery to regenerate. Virtually every human influence creates a series of events that impact the ecosystem services we rely on. Such is nature.

Let's survey the Earth's land, air, and water to explore a few examples of how humans are influencing natural ecosystems. The recurring theme is that virtually all of Earth's ecosystems have been transformed through human actions.

Land: From Natural to Artificial Ecosystems

If humans weren't around, the continents would hold continuous wilderness from ocean to ocean. One ecosystem would fade into the next, and some would occupy massive areas. Rich diversity and large mammals would be common throughout. But humans *are* around.

The United States has a total of 2.3 billion acres. According to the U.S. Department of Agriculture, about 52 percent of the total is used for agriculture, 23 percent for commercial-use forests (i.e., timber production), 7 percent for residential areas, 1 percent for military bases, and 1 percent for roads. A network of roads penetrates every corner of the country, including the national parks. Only about 11 percent of land is considered natural area for recreational and/or preservation purposes. Even the "natural" areas in much of the country are actually developed land that was abandoned several decades ago. The point is not that there are no significant natural areas in the United States, because clearly there are (nearly a third of U.S. land area is forest, for example). Instead, the point is that virtually all land has been significantly impacted by humans in one way or another. If we were to define "nature" as an area untouched by humans, then hardly anyone alive has actually seen nature. In this regard, the United States is a microcosm of the entire planet and, in fact, land use statistics for the world as a whole are similar to those above.

Humans are also shaping the topography of Earth, which transforms ecosystems. Most notably, Earth is being flattened. Geologists in the early 1800s were already noticing the "leveling" of the Earth by humans, as illustrated by Charles Lyell's *Principles of Geology* (1830):

> [Humans] convey upwards a certain quantity of materials from the bowels of the earth in mining operations; but, on the other hand, much rock is taken annually from the land, in the shape of ballast, and afterwards thrown into the sea, whereby, in spite of prohibitory laws, many harbours, in various parts of the world, have been blocked up. We rarely transport heavy materials to higher levels, and our pyramids and cities are chiefly constructed of stone brought down from more elevated situations. By ploughing up thousands of square miles, and exposing a surface for part of the year to the action of the elements, we assist the abrading force of rain, and destroy the

conservative effects of vegetation.

The scale of the flattening has increased as our technology for moving soil and rock around has increased. As has been the case for thousands of years now, erosion related to agriculture and deforestation washes an unimaginable amount of topsoil from the land and into waterways. Mountaintops are being removed in Appalachia and elsewhere for coal and other substances. Bulldozers and trucks constantly work to flatten ground for new roads, buildings, and parking lots. Most land has been cleared and plowed at some point for one purpose or another, whether in modern times or in centuries past.

All of this flattening has a homogenizing effect on ecosystems. The more varied a terrain, the more varied the habitats, and therefore the more biodiversity it will contain. For example, consider a hill in the Northern Hemisphere. Because the sun rises in the east, sets in the west, and occupies the southern sky for much of the year, there will be big differences in light and moisture at different locations around the hill. The southern side will get more sun and be drier, the northern side will be cooler and moister, water will collect at the bottom, wind will be more prevalent at the top, and so forth. As a result, you would often find differences in the species present at each of these locations on the hill. Flattening the hill, however, would reduce habitat diversity, which would reduce the biodiversity.

Although humans have destroyed a great number of terrestrial ecosystems and landforms, it would be simplistic to think that human actions have been entirely destructive. The reality is a much more complex reshuffling of nature. Consider the complex impacts of a relatively simple activity—purchasing soil for home gardening. The soil is dug up from one location, placed into plastic bags (which were made from oil taken from the ground in the Middle East), and shipped to another location. The soil is then added to someone's yard, where it blends with a completely new ecosystem, while the plastic bag makes its way to a hole in the ground somewhere else (a landfill). In most locations in the developed world, one cannot pick up a handful of topsoil or rock from the ground and know for sure where it came from.

Because different organisms thrive in different types of soils, all of this reshuffling of land has profoundly influenced ecosystems. Every bag of dirt exchanged from one location to another carries with it soil microbes from the previous site. Also, the types of plant species that occur at a given location are affected by soil acidity, particle size, moisture retention, nutrient content, and other factors. Changing any one of these factors in soils can tip the competitive balance such that different species begin to dominate and out-compete others, thereby changing the dynamics of the whole ecosystem. For example, adding nitrogen to an ecosystem often leads to rapid ecological change, as anyone who fertilizes a lawn can observe (because different plant species thrive in different nutrient conditions).

Overall, humans have transformed most terrestrial ecosystems, either by developing and then abandoning them, or by actively maintaining artificial ecosystems like cities and farmland. There are few places left on land that have not been cleared, plowed, dug, burned, or otherwise changed by humans.

Atmosphere: Changing Global Climates

Our use of energy is among the best examples of human influence and dependence on the rest of nature. When you pump fossil fuels into a car, you are essentially pumping liquefied dead organisms from millions of years ago. More specifically, most oil derives from unicellular plants (phytoplankton) that lived in the oceans during the Jurassic period. These organisms used photosynthesis to pull carbon from the atmosphere and convert it into sugars. After they died, they fell to the ocean floor and were buried. After being cooked deep in the Earth for millions of years, the dead remains became oil. Modern humans drill into the Earth, extract the oil, burn it, and release a huge amount of carbon dioxide into the atmosphere. Oil is used in the production of a huge number of things in our society, including roads, plastics, fabrics, fertilizer, medicines, and toothpaste. And those items not made directly from oil are probably transported using oil.

Burning oil, coal, and other fossil fuels releases carbon dioxide along with other greenhouse gases into the atmosphere. These gases serve as a thermal blanket for the Earth. After sunlight penetrates the atmosphere, some of the energy is held in the form of heat and some of it is radiated back into space. Adding extra greenhouse gases traps additional heat, adding additional energy to the climate systems. This phenomenon is often referred to as "global warming," but this is a poor descriptor for what is going on. "Climate change" or perhaps "global weirding" are more accurate in that everything related to climate is changing in ways that are often difficult to predict. Many regions of the world are experiencing increased average temperatures, whereas others are likely to see decreases. Some regions will see increased precipitation, whereas others will see far less. Since both agricultural systems and natural ecosystems are adapted to particular climates, changing climate creates major upheaval in nature and society. It is no coincidence that previous mass extinctions were related to climate changes.

Even if changing the chemical composition of the atmosphere did not alter climate, it would still have wide-ranging effects on ecosystems. Plants grow differently in different levels of carbon dioxide and ozone. Different species of plants respond differently — one might grow much faster in a new atmospheric environment, while another grows slightly slower. The result is that altering the chemical composition of the atmosphere gives some species a new competitive advantage. New rounds of natural selection take place, allowing some species to out-compete others and dominate ecosystems while others slide into oblivion. This is being documented by scientists performing FACE (free air CO_2 enrichment) experiments in ecosystems all around the world.

The people hurt the most by climate change are often those who did the least to cause it, namely those in the less industrialized, developing countries. Climate change reduces crop yield (or at best forces rapid agricultural adaptation) in regions of Sub-Saharan Africa, South America, and elsewhere, forcing people to migrate to city slums or other countries. Many of the recent migrants into the United States have also fled harsh environmental and agricultural conditions caused by climate change. The lesson that should be learned by industrialized countries is that when you contribute to poverty in other countries, they will soon be at your doorstep.

Climate change also damages food production in the oceans. Increased carbon dioxide levels in the atmosphere lead to more carbonic acid in the oceans, which increases acidity overall. Increased acid levels contribute to the bleaching effect that is destroying coral reefs around the world. Much of the world's population relies on the oceans for fish and other food, and much of the ocean's bounty is found on or near coral reefs. It has been estimated that about a third of reef-building corals face elevated extinction risks due to climate change, threatening the food supply of hundreds of millions of people.

Water: Industrial Fishing and Pollution

Because two thirds of the Earth's surface is water, the oceans might seem like inexhaustible resources. However, our use of the oceans has increased dramatically as human population has increased. Today, less than 5 percent of the oceans appears to be in pristine natural condition. Overfishing, sewage, climate change, pollution, coastal development, and disturbances associated with shipping have all stressed ocean ecosystems. Over 90 percent of ocean species harvested by humans have been depleted to less than half of their original populations.

In some cases, such as the massive oil spill by the Exxon Valdez off the Alaskan coast in 1989, ecosystems are severely damaged or destroyed very quickly. In other cases, different stresses have a long-term cumulative effect. We now understand that human impacts are driving changes to ocean ecosystems throughout the world. Among these are decreased numbers of large organisms (such as sea turtles, whales, and dolphins), decreased size of individuals within species, destruction of coral reefs and other ecosystem types, and the creation of artificial ecosystems (such as shipwrecks and developed coastlines).

One of the biggest impacts on ocean ecosystems has been trawling. Just as Earth's land has been scraped for agriculture, mining, and development, the oceans have also been scraped. Trawling involves dragging the sea floor to net fish, shrimp, and other organisms. Much of the Earth's oceans are trawled over and over every year, ripping corals, sponges, worms, and other organisms from the sea floor. Trawling also spews plumes of sediment into the water, where it can linger for hours. Like clear-cutting forests, trawling destroys habitat, decreasing the ability of ecosystems to replenish themselves. Commercial fisheries are moving their operations further and further offshore as coastal fisheries are depleted. The open seas are unregulated, allowing the bottom of most of the world's fisheries to be trawled and then abandoned.

Essentially every source of fresh water has been impacted by humans, and many have become highly engineered systems. The world's rivers have been rerouted and dammed, creating new lakes and destroying old ones. The Millennium Ecosystem Assessment (2005) estimated that construction has moderately or strongly affected flows of 60 percent of the large river systems in the world. Likewise, fresh water is often polluted by human and livestock waste, as well as all sorts of chemical residues from medicines, plastics, and other human products.

Among the biggest forces altering fresh water ecosystems are agriculture and climate change. About 70 percent of fresh water usage is for agriculture, usually in the form of crop irrigation. Climate change is altering rainfall patterns and melting glaciers, which provide fresh water for much of the world's population. Lake Chad is a stunning example of how agriculture and climate change can dramatically change a fresh water source. Once among the largest lakes in Africa, Lake Chad has now almost disappeared. It shrank more than 95 percent over 35 years due to irrigation projects and drought. As a result, a harsh region of the world has become even more inhospitable.

◆ ◆ ◆

In summary, human dependence on natural resources has led us to dramatically change the Earth's land, atmosphere, and water. As a result, we have reconfigured ecosystems all around the world and in many cases degraded their ability to provide food, materials, oxygen, water, and other ecosystem services. Achieving balance between the use and restoration of natural resources is essential to our future, because our ability to maintain the health of natural ecosystems is directly linked to social, economic, and political stability. As ecosystems go, so goes civilization.

Obviously, you can't change the planet's land,

atmosphere, and water without also changing the species that live there. The activities of civilization are driving evolution on a grand scale. In the next chapter, we explore what humans have done to the diversity of other species.

4

FORGOTTEN WORLDS

The Decline of Biodiversity

I have killed the bear and hyena, the lion and panther, the tiger, the stag and the ibex, all sorts of wild game and the small creatures of the pastures. I ate their flesh and I wore their skins.

— *The Epic of Gilgamesh* (3rd millenium B.C.), translated by N.K. Sandars

*I*t pays to be a "glass-half-full" kind of person. Optimism and hope drive us forward to better futures, while cynicism and negativity cloud our paths and misdirect us. Nevertheless, we have to accept responsibility for our mistakes before we can find better paths forward.

Our manipulation of Earth's land, atmosphere, and water has had some very negative evolutionary consequences. In particular, we have threatened millions of species with extinction. It took evolution more than 3 billion years to generate modern biodiversity, and we have imperiled much of it in less than 20,000 years. The evolutionary implications of our destructiveness are massive. Whole branches of the tree of life are under threat.

If we accept our responsibility as stewards of evolution, then preserving and restoring biodiversity must be a primary concern. We depend on biodiversity for ecological stability and adaptability, medicinal cures, disease resistance in agriculture, and new economic opportunities. Biodiversity also makes for a more beautiful world, improving our quality of life in ways that are difficult to quantify. And if we are humble in our view of life, then biodiversity is an end in itself. Surely we are not so selfish and brutish as to think that life should only exist to serve us.

Modern civilization is much like a giant tree house in the branches of the tree of life. Sometimes a small branch or two can be carefully pruned without major consequences. But if we start hacking away thoughtlessly, we damage the tree and endanger our tree house. Currently, we are hacking away at the tree of life like a drunken, ax-wielding lumberjack. We had better sober up before we chop off whole portions of the tree and topple parts of civilization along with it. Consider this chapter a little cold water in the face of the drunken lumberjack in all of us.

Two modern realities are especially sobering. First, much of the grandeur of wild nature has already been destroyed. Second, hardly anyone realizes the breathtaking extent of what we have already lost. As we consider the future, this second reality is especially problematic because it means that we're not learning from our mistakes. Let's consider what we've lost so that we can chart a better future.

Shifting Environmental Baselines

For an American, a good starting point is to imagine a parakeet. This particular parakeet is bright green, with patches of yellow, orange, and red. It is more colorful and more beautiful than the native North American birds that we know. If you were to see it in nature, it often would be traveling in a large flock. Imagine seeing dozens or even hundreds of these brightly colored birds sitting in a single tree. It would be an uncommonly beautiful spectacle.

Where would you expect to find such a parakeet? Most people imagine it living in a rainforest, perhaps in South America or Southeast Asia. Others will think about pet stores, Disney World, or some other artificial setting.

As you may have guessed, this is not a fictional bird, but a forgotten one. The bird's ecology made it very susceptible to human influences. It roosted in hollow trees, and so it required old-growth forests for breeding. It was also hunted for its colorful feathers to decorate hats and other objects. Farmers killed large numbers of them because the birds damaged crops, especially apples and other fruits. But its ultimate downfall may have been its social behavior. Apparently, it was unafraid of humans and was easy to approach. Then, after one was shot, others would circle the dead or wounded bird, allowing a hunter to kill an entire flock in a matter of minutes.

The bird was the Carolina parakeet (*Conuropsis carolinensis*), the only parrot native to the United States. It was once abundant throughout the entire eastern United States and, if humans had made different choices, it would still be here today. The last one died in captivity in 1918 in the Cincinnati Zoo. Over the course of several generations, humans drove the Carolina Parakeet from dominance to complete extinction.

Now the Carolina parakeet, like virtually every other extinct species, is eliminated from existence and most human memory. Why do people in the United States, Europe, and other developed countries associate large, colorful, and wondrous animals with other parts of the world? The unfortunate reality is that the most magnificent animals in their region have already been exterminated or driven into faraway refuges.

How can people be unaware of biodiversity loss in their own region? As Callum Roberts explains in *The Unnatural History of the Sea*, "When [environmental] baselines shift, each new generation subconsciously views as 'natural' the environment they remember from their youth. They compare subsequent changes against this 'baseline,' masking the true extent of environmental degradation, even to the degree that they no longer believe anecdotes of past abundance or size of species." In other words, most people have no idea about what they are missing. Sadly, this is even true for many biologists.

To understand our own obliviousness, it might be useful to imagine the life of future generations without parts of nature that we take for granted. For example, imagine generation after generation growing up without trees—no tree swings, no sitting under a tree on hot days, no climbing trees, no watching leaves drop in the fall and return in the spring, no paper, no fruit, no wooden furniture, no nuts, and no wildlife associated with trees. How could you explain the value of trees to people who never experienced them? Once you can imagine how oblivious such a person would be, then remember that from the perspective of past generations, you *are* that person.

Geologists and evolutionary biologists generally recognize five great mass extinctions in the past, times when the number of species dropped precipitously in a short period. The last one occurred 65 million years ago when the age of dinosaurs came to an abrupt end. Comparing modern trends with the past suggests that we are in a sixth great extinction. In other words, humans are causing a decline in biodiversity that is comparable to the greatest extinction events of all time.

Extinction rates in the modern world are about 1,000 times higher than the average extinction rate across geologic time. Let's look briefly at what we've already lost.

The Decline from 15,000 B.C. to A.D. 1500

It is difficult to imagine what North America was like when humans first arrived on the continent more than 15,000 years ago. When you think of elephants, big cats, and other large mammals, you probably think of Africa or Asia. The reality is that there were many large animals in North America not so long ago. These included woolly mammoths, mastodons, camels, saber-toothed cats, musk ox, tapirs, peccaries, giant ground sloths, buffaloes, red foxes, mountain lions, and grizzly bears. Along the coastlines were walruses, whales, penguins, dolphins, sea turtles, otters, seals, sea lions, and other large fish and aquatic mammals. Large animals were once everywhere in North America— hundreds of millions of them.

Early humans in North America probably saw large animals with the frequency that we see cars. If this doesn't seem possible, I invite you to visit the last refuges of the ancient wildernesses. Visit the Serengeti, the Ballestas Islands, or the Great Barrier Reef. These ecosystems were once typical, not extraordinary. In just one or two days in the Amazon rainforest, I have encountered more than 100 monkeys (squirrel, brown, capuchins, and saddle-back tamarins), jaguars, giant river otters, black caimans, dwarf caimans, dozens of turkey-sized birds called trumpeters, parrots and other colorful birds, large mice, brightly colored frogs, snakes, and many other animals.

Large animals are only the beginning of the biodiversity in any unspoiled wilderness. In mature rainforests, the plants are so diverse that even experienced field biologists struggle to identify many of them. Many trees have enormous bases and are frequently covered with other plants (called epiphytes), insects, and fungi. If you sit still for a while at the right time of day, the insects move around in such incredible numbers that it's like looking down on New York City traffic. Organic movements and sound are everywhere. Bedazzling shapes and colors are all around.

In North America, we don't know much about many species that early humans drove toward extinction, and we probably don't even know all the major species. Therefore, it is impossible to know exactly what these ecosystems were like. Nevertheless, they surely rivaled the most amazing ecological spectacles of the modern world. For example, in the center of North America all the general components of the modern African savannah seemed to have been present—vast expanses of grassland, huge herds of various hoofed animals, and large predators. The North American version might have been more spectacular.

The large animals began declining about the time humans began arriving in North America. Declines in large animals led to many other changes in ecosystems, if not total ecological transformations. Like today, the downfall of many species may have been partly due to climate change, but it was also driven by hunting and land use practices of early Americans.

The modern romantic idea that all Native Americans lived in harmony with nature is not supported by evidence. This idea results from comparing Native Americans with a culture that was even more destructive to biodiversity—the colonizing Europeans.

The Decline from 1500 to 2000

Since A.D. 1500, ecosystems around the world have changed dramatically. As a result, around 850 species went extinct between 1500 and 2000 that we know about. Probably many times that number went extinct without anyone noticing.

Some of the species that have vanished were among the most prolific and robust species on the planet. For example, passenger pigeons were once the most common bird in North America, but overhunting led their populations to collapse. The last one died in 1914. Eastern elks were once among the dominant hoofed animals in eastern North America. The last ones were seen in the 1870s. Heath hens were common birds of New England during the colonial period.

Despite efforts to prevent its extinction, the last one died in captivity in 1932. Sadly, the list goes on and on.

For every species that has gone extinct, many more have been marginalized and isolated. For example, the American buffalo (bison) numbered in the millions when Europeans first stepped foot in North America. It was official U.S. government policy to destroy the buffalo herds because many Indian tribes were dependent on them. After a continent-wide slaughter, there were only a few hundred buffalo left by 1900. Since then, populations have rebounded somewhat, but that is only counting buffalo raised on farms.

Likewise, the American chestnut tree was once among the most useful, abundant, and ecologically important trees in the eastern United States. Chestnuts were all but wiped out by an Asian bark fungus, an imported disease first noticed in the Bronx Zoo in 1904. It spread south over the next several decades. In total, about 3 billion chestnut trees were destroyed in the twentieth century.

Many species that appear healthy today are illusions in terms of biodiversity. Heavily hunted and trapped species were wiped out in many regions and then restocked using animals from another area. For example, North Carolina beavers were wiped out by trappers and later restocked using beavers from Pennsylvania. Deer, which are now overabundant in some parts of the Southern United States, were once decimated and restocked. Thus, significant biodiversity has been lost within species, even those that are abundant today, making them more susceptible to disease and changing environmental conditions.

The Current State of Biodiversity

According to the IUCN Red List (2010), there are currently more than 17,000 threatened or endangered species that we know about through rigorous scientific study. This includes at least 21 percent of mammals, 30 percent of amphibians, and 12 percent of birds (the three most studied groups). Species on the chopping block are tigers, gorillas, Caspian seals, Grevy's zebras, Tasmanian devils, Iberian lynxes, slender-billed vultures, ploughshare tortoises, European eels, purple marsh crabs, western prairie fringed orchids, and thousands more.

These IUCN data include only species we have studied, which is a tiny proportion of the total. How many more species might be threatened or endangered that we don't know about? Consider the following limitations of the IUCN data:

- Scientists have evaluated the status of only about 48,000 of the more than 1.7 million species recognized by the IUCN.
- Biologists estimate that there are between 5 and 30 million species on the planet, far more than are taken into account by IUCN data. Many species exist that have not been described by science.
- The data do not (and cannot) take into account the potential effects of climate change, which could rearrange ecosystems virtually everywhere.

Thus, the true number of threatened or endangered species probably reaches into the millions. The numbers increase annually as problems worsen and data collection improves. Some scientists have argued that more than half of the world's biodiversity could be gone by the end of the twenty-first century. There is no point in quibbling over the exact number. The number of threatened or endangered species is clearly enormous.

In the modern world, there are five main causes for decreasing biodiversity: habitat loss, invasive species, overharvesting, pollution, and climate change. First, habitat loss is the most obvious and the most destructive. Essentially every ecosystem on the planet has been modified within the past few centuries. Currently, the world's most concentrated sources of biodiversity are being threatened. For example, the tropical rainforests are being cut down at an alarming rate. Likewise, trawling for fish is destroying the deep oceans often before anyone looks to see what is down there. If organisms do not have a habitat, they will not survive.

Second, invasive species are non-native organisms that are introduced to an ecosystem and take over. Some of the most notorious invasive species were introduced intentionally by humans, who badly misjudged the effects they would have. In many other cases, global transportation and commerce have led to unintentional mixing of species. Seeds, insects, worms, and microbes hitch rides in cargo holds, pants legs, airplane landing gear, and every other imaginable place as they travel the globe. Invasive species account for more than 25 percent of the species in many "natural" ecosystems. The more notorious invasive species in the United States include kudzu, Asian tiger mosquitoes, hemlock woolly adelgids, Japanese beetles, and zebra mussels. Other invasive species have been around so long that most people think they are native (e.g., various species of earthworms and dandelions). The U.S. Department of Agriculture has estimated that over 18 million acres are currently occupied by invasive species. It goes the other way, too—native American species such as bullfrogs, gray squirrels, Leidy's comb jelly, and largemouth bass serve as invasives in other parts of the world.

Third, overharvesting is perhaps the most primitive form of biodiversity loss, as it has contributed to human-driven extinctions time and time again for thousands of years. Examples in recent history abound—Carolina parakeets, dodos, great auks, Bali tigers, sea minks, passenger pigeons, and Caribbean monk seals. By the late 1800s, almost every commonly hunted mammal and bird in the United States was in decline (some have since recovered, others have disappeared). A more recent example of overharvesting is the collapse of cod fisheries on the Atlantic coast of North America in the early 1990s. For decades, the fisheries yielded huge catches of cod every year, supporting a large fishing industry. After trawling was introduced, fish catches temporarily spiked as fishermen caught unprecedented numbers of cod. But trawling involves scraping the ocean floor and netting everything that is kicked up, ripping apart the ocean floor and destroying habitat for fish and shellfish. Soon fish populations crashed, hardly any cod could be found, and the fishery closed to cod fishing. Many thought that cod were so vigorous that their populations would bounce right back, but the fishery has never recovered.

Fourth, pollution can take many forms, from

dumping raw sewage into waterways to pumping greenhouse gases into the atmosphere by cars, power plants, and factories. A difficulty in preventing pollution is that the polluters are often not the people who suffer the most immediate consequences. The current situation in the Gulf of Mexico is a classic example. Around the coastline of Louisiana, where the Mississippi River empties into the ocean, there is a growing biological dead zone where no organisms can live (except for microbes). The scale of the dead zone is massive, occupying over 6,000 square miles of ocean. Since the Gulf of Mexico supplies a large proportion of the shrimp and oysters for the United States, this is a major threat to the ecology and economy of the region. To solve the problem, though, one would have to improve the agricultural practices of Midwestern farmers, who are hundreds of miles away. Farmers in the Midwest dump huge amounts of fertilizer, herbicides, and pesticides onto their fields, which wash into local waterways and end up in the Mississippi River. This agricultural run-off is coupled with other pollutants dumped along the long trek to the ocean. The Gulf of Mexico has basically turned into a toilet for Midwesterners and the fossil fuel industry. This epitomizes pollution problems globally—many people prefer that their toilet be where someone else lives.

Finally, climate change is the great biodiversity wildcard. We know that changing climate will force massive shifts in ecosystems worldwide, and we know that many species will not survive these shifts. We also know that similar climate shifts have contributed to mass extinctions in the past, so there is good reason to be worried. When ecosystems change extremely rapidly, certain types of organisms are at a huge evolutionary disadvantage. Specialist species may be unable to adapt quickly enough to a new ecological niche before they are totally wiped out. Modern climate change could be much more disastrous for biodiversity than were similar changes in the past because natural ecosystems have been carved into many discontinuous pieces. During past climate changes, it was easier for organisms to migrate from place to place, moving with climate regions and adapting with global change. This is still possible for some species such as birds and plants with wind-dispersed seed. But for many others, human development has isolated them on small islands and, without assistance, climate change is a death sentence.

Then there are all of the smaller impacts that human societies have on biodiversity. For example, domesticated cats are popular in many regions of the world. Studies have shown that collectively, cats kill billions of small mammals and birds every year. Another example is bird death due to crashing into windows. Birds can't see most windows and, just like the occasional human who walks into a glass door, they crash head first and break their necks.

And how many animals are killed by cars and the presence of roads? I wrecked my first car by hitting a bear who was crossing a highway at night. Individually, each event may seem inconsequential or trivial. But taken together, human activities are causing a massive biodiversity decline that is changing the course of evolution.

The Successful Few

As Julian Huxley observes in *Evolution: The Modern Synthesis*, "It seems to be a general characteristic of evolution that in each epoch a minority of stocks give rise to the majority in the next phase, while conversely the majority of the rest become extinguished or are reduced in numbers." Humans have dramatically accelerated this process by driving biodiversity decline and homogenization. Based on the selective pressures applied by humans, which species are the most likely to survive the sixth great extinction and proliferate over the next million years? Let's consider two groups of potential survivors, which I will call the "anthroadaptors" and the "anthrochosen."

The Carolina parakeet epitomized the traits that have proven maladaptive in a world full of humans — they were beautiful, social, and required old-growth forests. By killing them and everything like them, we unintentionally select for the "anthroadaptors," organisms that can adapt to human development. Generally speaking, the characteristics of the anthroadaptors include:

- Being a generalist (tolerance to a wide range of environments)
- Lack of herding or flocking behavior
- Small size
- Skittish behavior around humans
- High mobility (especially for climate change adaptation)
- Unattractive appearance (from the human perspective)
- Bad taste (from the human perspective)

• Rapid reproductive rate

By hunting, harvesting, and exploiting, humans select against the traits that make species valuable to us. Thus, over time, we are left with organisms in natural ecosystems that we find less valuable. The ultimate anthroadaptors are the species that thrive around intense human development—mice, rats, squirrels, raccoons, pigeons, opossums, house flies, common yard weeds, house molds, termites, ants, tiny fish that slip through our nets, and viruses and bacteria living on and around humans. Most of these species are considered nuisances. They are characterized by their uselessness to humans and the cleverness of their survival strategies. They are too small and scattered to be worth hunting, too ugly to have as pets or ornamentals, and too prolific to wipe out.

When you think about it, you can't help but respect these humble creatures. They survive our domination of the planet without becoming our playthings. If we destroy ourselves, they will quickly evolve to fill every available niche, repopulating the Earth with new types of large species. They have much in common with the small species that avoided the dinosaurs, survived the last mass extinction, and eventually evolved into humans.

The anthrochosen are the organisms that humans intentionally choose to live among us. These include our crops, livestock, ornamental plants, and domesticated pets. Anthrochosen species are selected for a different set of characteristics, depending on their purpose:

- Gentle behavior
- Small size (e.g., pets and shrubbery) or medium size (e.g., livestock and landscape trees)
- Rapid growth (e.g., crops and livestock)
- Beauty
- Homogeny

The last two characteristics are the result of our peculiar human biases. We have a particular idea about what an apple, a rose, or a cow *should* look like, and we are generally intolerant of diversity. Anthrochosen species are selected to fit our stereotypes. We have taken them in as our evolutionary companions. If we survive, so do they. But they have taken us in as well. In fact, we have built a civilization that cannot exist without the anthrochosen.

What about the millions of other species that do not fit into either the anthroadaptors or anthrochosen as we've described them so far? There are two options at this point. Either we proceed with the current momentum of civilization and many of them will cease to exist, or we change our priorities and decide to broaden our definition of the anthrochosen. As long as billions of humans are around, most species can survive *only* through a conscious decision-making process by humans. We either conserve natural ecosystems, or we do not. Taking no action to conserve species leads to the same result as actively trying to destroy them. And it leads there very quickly.

In the meantime, as we contemplate what we've already lost, we may wonder about how our choices will impact the lives and experiences of our descendants.   C.S. Lewis vividly paints a possible future in "The Future of Forestry":

> Simplest tales will then bewilder
> The questioning children, 'What was a chestnut?
> Say what it means to climb a Beanstalk.
> Tell me, Grandfather, what an elm is.
> What was Autumn? They never taught us.'

It doesn't have to be this way, of course. The future is a choice, not a destiny. But do we have the wisdom to learn from our mistakes? Can we find a balance between what we take from nature and what we put back?

In the next chapter, we consider the concept of sustainability and explore our options for the future.

5

NATURE OR NOTHING

Sustaining Ecosystems and Societies

Men and nature must work hand in hand. The throwing out of balance of the resources of nature throws out of balance also the lives of men.

— Franklin Delano Roosevelt
(message to the U.S. Congress, January 24, 1935)

*L*ooking back at history, it is hard to imagine that people could believe some of the things they did—the Earth is flat, life spontaneously generates itself, and disease is caused by evil spirits. These ideas seem so ridiculous to the modern, educated mind that it can be difficult to appreciate the cultural context for them. Future generations will add to this list the relatively modern idea that the Earth is too big to be fundamentally changed by humans. Not only can we fundamentally change it, we already have. Now we find ourselves struggling to maintain the world that sustains us so well, the world into which we evolved.

Sustainability refers to the ability to live without decreasing the ability of future generations to meet their needs. On the surface, it doesn't sound too difficult. After all, we're the result of an evolutionary process that took over 3 billion years, accounting for millions and millions of generations. We all made it here, so we have proved the concept—sustainability is possible. On the other hand, the modern world is different. The human species is relatively young, and we are the first species to gain so much control over the evolutionary process. It is an open question whether we have the wisdom and self-control to develop a "culture of permanence," as E.O. Wilson puts it.

The definition of sustainability above is selfish in that it refers exclusively to the future of humans. It is primarily *our* future that concerns most of us. Although many people value other species for their own sake, such sentiment is often fleeting when what helps another species seems in conflict with what helps humanity. Likewise, there are some people who care little for natural ecosystems and, if they had the power, would have already destroyed them. Thus, as we consider our evolutionary options for the future, we can imagine a wide range of possibilities, from a totally engineered world to a world of restored natural ecosystems. Consider these generalized possibilities for the future:

- **Level 1**: Preserve no natural ecosystems. This would require that eventually we rely entirely on engineering to provide clean water, air, and food.

- **Level 2**: Minimize ecosystem degradation. Allow ecosystems to be further degraded by prioritizing development as most important, but keep some environmental laws in place to slow down the damage. Level 2 best represents the current policies of most countries.

- **Level 3**: Preserve the current state of ecosystems. Concede that what is lost is lost, and focus on maintaining everything that is left. Tightly control development and expand parks, preserves, and other natural areas.

- **Level 4**: Preserve current ecosystems and restore others. At least 15–20 percent of land and oceans would be permanently protected in parks and preserves. The development of any natural area would require restoration somewhere else. The ultimate goal is to slowly rebuild the stature of natural ecosystems over time.

- **Level 5**: Return large areas to natural ecosystems. At least 20–50 percent of land and oceans would be permanently protected so that large animals could roam freely. Much of the grandeur of true wilderness would be restored over time.

The choices span a wide range of outcomes for the future of life. What should we choose? In the long term, should our ultimate goal be to maintain a few green patches in the landscape, preserve ecosystems in their current degraded form, or truly restore the landscape? Let's consider our evolutionary options in more depth.

Level 1: Preserve No Natural Ecosystems.

We already made note of the services that ecosystems provide to humans, including clean water, breathable air, and food. But we can clean water using water treatment plants. We can grow food in agricultural settings. Is it possible that we could engineer the whole planet so that human ingenuity replaces natural processes?

Natural ecosystems consist of hundreds of interconnected cycles and relationships. Water falls from the sky, nourishes plant life, runs into rivers and then the sea, and eventually evaporates back into the sky. Plants take in carbon dioxide and release oxygen; animals do the opposite. Ecosystems also cycle nitrogen, phosphorous, carbon, and other elements. Likewise, each species represents an interconnected cycling unit. For example, bees have their own life cycle, and aid the lifecycles of plants through pollination. Ecosystems manage to cycle and "re-cycle" everything. To exist without them, then, we have to figure out how to recycle everything. It would be unfathomably complicated.

There have been many attempts to create self-contained ecosystems that could sustain human life. If successful, these could offer evidence that perhaps the whole Earth could be sustainable without natural ecosystems. For example, space shuttle flights and space stations are examples of totally closed systems. However, these are not currently sustainable because food, fuel, and supplies must be continuously added to sustain humans for any length of time. Every country interested in space exploration has attempted closed, sustainable systems, including NASA's BioHome, the European Space Agency's MELiSSA, and the Soviet

Union's Biosphere 3. Another popular example is the Biosphere 2 experiment in Arizona, an enormous closed dome covering more than three acres. Each set of experiments yielded interesting results from a scientific standpoint, but none proved to be sustainable. The humans in them were pulled out after a few weeks or months when the air or water became toxic. In fact, humans have never built a sustainable, self-contained ecosystem in which they can live. Let's face it—as engineers of sustainable ecosystems, we stink. At least for now. If we severely damage global ecology, we do not yet have the knowledge or means to fix it.

Some will wonder if these experiments are relevant for a system as big as Earth. Indeed, it is commonly assumed that the Earth could never be made uninhabitable for humans. The planet is too big and we are too small, the argument goes. What would be the evidence for this? Are we indestructible? Both time and space offer definitive answers. First, time tells us that 99 percent of species on Earth have already gone extinct. Second, space tells us that there are billions of stars, many with many planets and moons associated with them, yet we have no hard proof that life currently exists anywhere but our own planet. The typical planet or moon is a dead, lifeless rock or a ball of toxic gas as far as humans are concerned. We can't calculate the odds at this point, but suffice it to say that uninhabitable worlds far outnumber habitable ones—perhaps a million to one (without changing them, that is). All evidence suggests that life in the universe is rare. Uninhabitable planets are the rule, and Earth is an exception.

By all accounts, the most scientific answer to the question, "Can humans exist without natural ecosystems?" must be "Not at this time." It is theoretically possible, but we are nowhere near being able to sustain human life without natural ecosystems. Thus, at this point, destroying them entirely amounts to societal suicide.

Level 2: Minimize Ecosystem Degradation

Most modern societies fall within the Level 2 category. In a Level 2 situation, there is some protection for natural ecosystems and concern for maintaining free environmental services, but short-term economic development is normally the first priority. Ecosystems are slowly degraded over time such that many people simply don't realize the extent of the damage.

The ultimate problem with Level 2 is that we will eventually degrade the planet to the point of having no more natural ecosystems (Level 1). It will just take longer. Eventually societies reach a point of runaway environmental destruction where the last natural resources become so valuable that it is almost impossible to prevent people from plundering them.

If we continue down the path of Level 2, then a number of more extreme conservation measures will become more popular. Specifically, biologists will attempt to "bank" biodiversity in zoos, cryopreservation chambers, and seed banks. The basic idea is that if we destroy much of the planet's ecology in the next century, perhaps future generations could rebuild everything using biodiversity in the banks. Such efforts are already underway. Some, such as the Global Seed Vault in Norway, are buried hundreds of feet underground and have security systems akin to Fort Knox. The thought of Earth's biodiversity existing only in zoos and freezers is sobering.

Zoos, cryopreservation, and seed banks are preservation tools of last resort. Although they are already valuable for small-scale conservation purposes, it is implausible that such efforts can maintain most species for more than a few decades. If it comes to actually relying on these efforts, we are in big trouble. First, many organisms do not maintain sustainable populations in such situations. Elephants in zoos, for example, usually live less than half as long as they do in the wild and have more reproductive problems. Only by periodically importing elephants into zoos are populations maintained over time. Likewise, cryopreservation assumes that in the future we will have technology that doesn't currently exist for bringing back organisms from only a few cells. Although such technology is likely and probably will "resurrect" key species, it is unlikely that we'll be able to customize it for thousands of different endangered or extinct species.

Even if the organisms in banks and zoos last, and we are able to bring back individual organisms, we would still have very little idea about how to restore Earth's ecosystems. To work in the long-term, future generations would have to be much smarter than we are in terms of ecology. If we destroy natural ecosystems, how can future generations better understand how they work? They would have nothing to study.

Level 3: Preserve the current state of ecosystems.

We have established that having natural ecosystems is essential for sustaining people. Thus, the possibility of long-term sustainability starts at Level 3 (but is still not guaranteed). Level 3 involves using the current state of ecosystems as the target for what we should maintain in the future. It suggests no further ecosystem degradation and equilibria between ecosystems and development.

Achieving Level 3 requires that a great deal more of Earth's surface, both land and water, would be set aside in parks and preserves. Most ecologists would agree that somewhere between 10 percent and 40 percent of Earth's land and oceans need to be set aside to maintain a sustainable global situation over the long run. In the United States, 11 percent of land is currently set aside in parks or preserves. To ensure Level 3 in the United States, protected areas would have to be greatly expanded. In Brazil, about half of rainforests exist within protected areas. Although a noteworthy achievement, this still leaves half unprotected. Level 3 would be most difficult to achieve in the oceans, because less than 1 percent of oceans are managed "no-take" zones.

To establish an equilibrium where no further ecosystems are lost, all development that destroys ecosystems would have to be balanced with ecological restoration somewhere else. Thus, monitoring and management would be important even for ecosystems that are not part of a park system.

There will surely be opposing viewpoints on how to achieve Level 3, because deciding exactly how and what to preserve is open to interpretation. As we've already seen, most ecosystems have been completely transformed over the course of thousands of years. Most efforts at preservation are, in reality, efforts to preserve ecosystems that are already highly degraded or fundamentally different from what they were before human intervention. Likewise, restoration targets usually involve restoring ecosystems that existed at a site within the last few hundred years, and rarely involve restoring prehuman ecology or anything with ecological grandeur. Level 3 maintains the modern extent of ecosystems, but it stops short of rebuilding what has been lost.

Level 4: Preserve Current Ecosystems and Restore Others.

Like Level 3, Level 4 preserves current ecosystems. But Level 4 would also require restoring natural ecosystems whenever possible and actually improving the environment. Level 4 suggests that we want to rediscover some of what has been lost. Not only do we want to maintain biodiversity and free ecosystem services, we want to increase them so that future generations have a better world to live in. Most ecologists would say that sustainability is assured once we enter Level 4.

In a Level 4 world, many ecosystems that have been fragmented throughout the landscape could be pieced back together by small and medium-scale restoration projects. This strategic approach would strengthen ecosystems, increase biodiversity, and allow larger animals to reestablish themselves in some regions.

As we consider the damage that has already been done to global ecology, it is easy to be pessimistic about the likelihood of actually restoring ecosystems on a broad scale. Nevertheless, examples of successful restoration projects are becoming more common. Very large areas of Puerto Rico, Tanzania, Costa Rica, Brazil, and other countries have been allowed to regenerate naturally. Also, nearly a thousand species have been reintroduced to the wild. For example, scientists in New

Zealand have successfully reintroduced plants, reptiles, invertebrates, and birds to natural environments. The four largest carnivores of western Europe (wolves, bears, lynxes, and wolverines), which were nearly exterminated by the nineteenth century, are making a comeback due to reintroductions and migrations. Gray wolf populations in the United States are rebounding. The U.S. Fish and Wildlife Service, which intentionally wiped out wolves a century ago, is now working to restore them in places like Yellowstone National Park. Perhaps the most iconic restoration in the United States is the bald eagle, which rebounded after the pollutant (DDT) that weakened its egg shells was banned. There are many examples to show that Level 4 is challenging but possible.

Level 5: Return Large Areas to Natural Ecosystems

Imagine a world where modern technological societies thrive, yet we maintain the grandeur of true wilderness. Level 5 suggests a transformational shift in how we interact with the planet. It suggests that not only do we want a sustainable world for humans, but we are also willing to share it to some degree with other organisms for their own sake. It would require a higher philosophy of civilization, such as Albert Schweitzer's idea that "A man is truly ethical only when he obeys the compulsion to help all life which he is able to assist, and shrinks from injuring anything that lives. He does not ask how far this or that life deserves one's sympathy as being valuable, nor, beyond that, whether and to what degree it is capable of feeling. Life as such is sacred to him." To the Level 5 advocate, life that is sacred includes other species as well as future generations of humans.

There is no historical example of a human society choosing Level 5. Only in instances of societies collapsing have Level 5 situations emerged as nature reclaims the landscape. Examples abound throughout the cultural centers of the ancient world. For example, huge tracts of Amazonian rainforest regrew after ancient South American cultures fell apart. In many cases where societies collapsed because of environmental problems, their Level 1 or 2 mentality eventually destroyed them, at which point Level 5 was thrust on them. Had they chosen Level 3, 4, or 5 voluntarily, they might still be around.

In recent years, restoration projects have become more frequent and more ambitious. For example, a massive reforestation project in the São Paulo region of Brazil is attempting to reforest 2.5 million acres. If such efforts become more common, then Level 5 could actually become a reality.

Principles of Sustainability

Humans are capable of accomplishing any of the aforementioned levels, from total ecosystem destruction to continent-scale restoration. Different outcomes will unfold over time as different countries and communities make their own decisions. It is likely that the societies that can sustain their natural ecosystems will survive, whereas others will end up desperately poor and unstable.

How can we design a society that is sustainable? Because technology and culture are different around the world and across time, sustainability takes a variety of forms. Nevertheless, there are three principles of sustainability that are relevant in any context.

The first principle is to preserve and restore natural ecosystems. If we wish to have natural ecosystems around to provide free services and enhance our lives, they must be preserved and restored. This includes establishing more protected natural areas, protecting threatened and endangered species, minimizing damage by invasive species, minimizing pollution and climate change, and eliminating overharvesting. Preservation and restoration serve the long-term interests of everyone, including the businesses that harvest natural resources. For example, a sustainable pulp and paper company plants as many trees as they harvest so that they will still have a business in the future. A sustainable fishing industry allows fish populations to regenerate, and a sustainable farmer builds soil and conserves water.

The second principle is to be more efficient. Maximizing efficiency includes minimizing waste, recycling, and using less energy, water, and materials. This once was just basic common sense—save your money and don't use what you don't need. For many of us today, however, excess and gluttony have become a standard part of life. As a result, we see economies struggling under mountains of debt and ecologies struggling under mountains of pollution. Thriving twenty-first century economies include a wealth of industries that are not based on selling excess "stuff," such as finance, tourism, education, media, and medicine. Likewise, a thriving modern economy includes many other industries that sell renewable products and renewable energies. The idea that a thriving modern economy must be built on buying and selling an excessive amount of disposable stuff is a myth.

We live in the Information Age, not the Industrial Revolution.

The third principle is to think, plan, and act for the long term. It is no exaggeration to say that the future of life on Earth hinges on this simple principle. A great deal of harm is done by people thinking from one quarter to the next, one election cycle to the next, or from paycheck to paycheck. Long-term thinking includes producing and purchasing energy-efficient and durable products, restoring the resources we consume, and supporting scientific research and education. This principle also brings the first two principles into proper context. For example, the idea that energy-efficient products cost more is a myth based on short-term thinking. Energy efficiency is nearly always more cost-effective in the long run, and so a savvy buyer will balance long-term and short-term costs and rewards. Likewise, the idea that draining a natural resource is "good for the economy" is absurd from a long-term perspective. It may be good for a quarterly profit, but it can be devastating to an overall economy over the course of a generation. Even children can understand the parable of the chicken that laid golden eggs. If we are greedy and kill the chicken in an attempt to harvest everything at once, then we are left with nothing.

On the surface, each principle is simple to understand. When it comes to commitment, though, all three have been extremely difficult for modern societies.

In the end, sustainability is based on the premise that we care about future generations. It suggests that our grandchildren, and their grandchildren, have the right to experience at least the same quality of life as we have. History shows that this is not to be taken for granted, as quality of life ebbs and flows over time.

Sustainable societies learn to coexist with their ecosystems, whereas unsustainable societies commit the slow suicide of ecosystem destruction. What future will we choose?

PART III
CELLULAR CHANGE

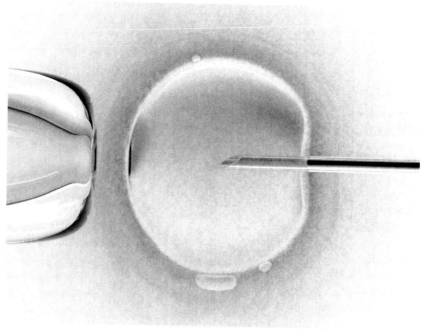

6

SIZE MATTERS

Manipulating Growth

> Our oldest cultivated plants, such as wheat,
> still often yield new varieties: our oldest
> domesticated animals are still capable of rapid
> improvement or modification.

— Charles Darwin (*Origin of Species*, 1859)

Consider your normal morning routine. You get up, take a shower, shave (if you're a guy), apply a bandage to that nick on your neck, get dressed, eat breakfast, brush your teeth, and so forth. Now consider the routine for what it really is—the daily management of cells. You wash to decrease the number of bacterial cells on your skin, shave to remove the protein deposits from your skin cells (otherwise known as hair), and apply a bandage to help skin cells replace themselves without being invaded by foreign cells. Then you take the cells from cotton plants and others organisms and drape them over your body to protect your own cells, and you eat the cells of other organisms to nourish your cells. Finally, you brush your teeth to decrease the number of bacterial cells in your mouth. When you think about it, we spend a huge proportion of our time managing and manipulating the growth of cells.

There are good reasons for our attentiveness to cells. Most notably, we are terribly outnumbered by foreign cells on and in our bodies. The average person reading this has more than 700 species of microbes living on a square inch of skin.

The diversity on your body is distributed similar to the patterns that govern the geographic diversity of larger organisms. On the dry desert that is your back, there are relatively few species. But life abounds in the warm, moist rainforest that is your armpit.

Inside our bodies, the situation is even more surprising. A typical adult human has 10 trillion human cells and 100 trillion microbe cells. That is not a typo – microbes outnumber us ten to one in our own bodies (this is possible because their cells are much smaller than ours). Although the numbers are enough to make us squirm, we shouldn't worry ourselves too much. Most of these critters don't harm us, and some even help as long as we are hygienic enough to keep our microbial "zoos" under control.

Pause for a moment to appreciate the implications of these numbers. Although they do not change *who* we are, they should heighten our self-awareness of *what* we are. We are part of the biosphere, and it is part of us. We are arks of cellular diversity. We are human, and we are also bacteria. And we are great manipulators of cell growth.

Growth and Cell Division

The word "growth" refers to a permanent increase in size. This might be an increase in population size for unicellular organisms, or it might be an increase in body size (or population size) for multicellular organisms. In both cases, growth is due mostly to cell division and not increased cell size. In other words, a big person has more cells than a small person, but there probably isn't much difference in the size of their cells. Thus, to understand growth, you have to know a little about cell division.

In unicellular organisms such as bacteria, cell division serves the purpose of reproduction. One organism divides into two, which divide into four, and so on to increase population size. In multicellular organisms, cell division increases an organism's size and aids in healing. This type of cell division, called mitosis, involves one cell cloning itself to form two cells. Multicellular organisms also have a second type of cell division, called meiosis, which produces sperm and egg cells for reproduction.

Because growth is determined largely by cell division, the amount of growth is determined largely by when cells "decide" to divide. How do they make such decisions? Biologists recognize four major stages in the cell division cycle: Growth 1 (G_1 phase), DNA synthesis (S phase), Growth 2 (G_2 phase), and Mitosis (M phase). This is a formal way of saying that a cell gets bigger, synthesizes new DNA, gets a little bigger again, and then divides. There are three "checkpoints" along this process where a molecular decision-making process takes place (at the end of G_1 and G_2 phases, and during mitosis). At each checkpoint, a cell responds to chemicals in its local environment. If conditions are favorable, it will continue with cell division. If conditions are not favorable, it will stop.

The checkpoint system is immensely important. If it works, organisms grow and repair tissue in a controlled and orderly manner. But if checkpoints are too rigid, growth and healing may be slow. Likewise, if checkpoints are too lax, cell division can outstrip available resources and endanger an organism.

If the checkpoints break down, uncontrolled growth causes tumors and cancers. As Bruce Alberts explains, "Cancer arises when the descendants of just one of our more than ten thousand billion cells proliferate out of control, eventually interfering with normal body functions. Since so many cells are at risk, the most amazing thing to me is how many years it usually takes to develop the disease."

Humans have been ignorant of the cell division cycle for most of the time that we've been manipulating growth. Nevertheless, it is cell division that we've been manipulating. For modern scientists who are further modifying organisms and trying to cure diseases such as cancer, it is cell division that we're still struggling to understand.

Intentional Manipulation of Growth

In previous chapters, we found that species of large animals have been selected against by humans, and in many cases have been driven to extinction. But that is only part of the story of how we are manipulating size in more natural settings. We are also changing growth and size *within* species. Through selective breeding and other means, almost every organism that we closely interact with has been significantly modified by humans.

Crop plants are a classic example of our manipulation of growth. Corn was bred from a grass called teosinte, a bushy plant with only a few small seeds. Over the course of several thousand years, humans changed it to a tall stalk with cobs over a foot long. Teosinte and corn now look nothing alike. Modern breeders continue to select plants with higher yield, disease resistance, drought tolerance, and other traits. Likewise, farmers improve the growth of corn by spraying herbicides on weeds and pesticides on insects, and by irrigating and fertilizing. Corn is a microcosm of all our crops. Rice, wheat, bananas, apples, tomatoes, and just about all the other major crops have been selectively bred from some ancestral plant and then carefully manipulated to maximize productivity.

Livestock have also been heavily modified over thousands of years to provide milk, meat, and fiber for humans. For example, cows were domesticated over 8,000 years ago. They were originally bred to be smaller and more passive, making them easier to control. Now, however, breeders have increased their size, and the mass was added to the parts that humans most want to consume. Dairy cows have udders the size of large beach balls, and beef cows have bulging muscles and fat. Growth hormones, antibiotics, and diet also increase the growth rate of cows. Remarkably, the rate of growth in beef cows more than tripled in the United States during the twentieth century, resulting in cows weighing over 1,000 pounds after just one year. Pigs, goats, horses, sheep, and other animals have their own unique histories of manipulation by humans.

The growth of our pets has also been manipulated. Dogs, for example, were bred from wolves over 12,000 years ago. Early humans surely valued their ability to hunt, defend territory, and serve as an alarm system. Some modern dogs keep wolf-like characteristics and are even intentionally interbred with wolves every few generations. The majority of dogs, however, have been modified to look and act dramatically different. The American Kennel Club recognizes more than 150 official dog breeds, ranging from tiny chihuahuas to massive Great Danes.

Most people would be shocked at just how intentional and calculated modern breeders are when choosing traits for future generations. They select for the precise size of virtually every body part, as well as texture, color, and behavior.

Other domesticated animals are manipulated in similar ways, each within their own unique constraints. For example, cats are meticulously bred, but there is good reason why domestic cats are kept as small as they are – big cats are top predators. If our pet cats were much larger, they might try snacking on our children from time to time.

Thus, most crops, livestock, and pets have been selectively bred to the point that they are often very different from their wild relatives. We have intentionally manipulated growth for health, convenience, productivity, beauty, and safety. This being the case, one wonders how we might manipulate the growth of crops, livestock, and pets in the future. How much yield increase is possible in agriculture? How fast could one grow an adult cow or pig? And how big is possible? Blue whales can be over 100 feet long and weigh over 170 metric tons, whereas the average brontosaurus was over 70 feet long and more than 20 metric tons, but these organisms also have (or had) body plans that allow them to be that big. The body plan of a species is eventually a limit to changes in size unless aspects of the body plan are altered. For example, rice plants were bred to bear more grains until eventually the plants started flopping over from the extra weight. Then breeders selected for a shorter stem that could bear the weight. Even with a redesigned body plan, the practical elements of managing a farm with giant animals might not be worth the effort. (Although it is humorous to imagine. . . "Someone go milk the brontosaurus-sized cow!")

Another future possibility is that there will be no livestock at all. Some scientists are trying to grow meat in the lab without whole animals ever being born and raised. Imagine having hamburgers without having to raise and slaughter cows. Is this more ethical, or less?

Whatever we decide to do, we must keep in mind that our choices eventually affect our own evolution. As Jared Diamond points out in *Guns, Germs, and Steel*, "We're just one of thousands of animal species that unconsciously 'domesticate' plants." The plants "domesticate" the animals in return as they become co-dependent on each other. Thus, we should expect that our bodies will adapt over generations in response to however we're modifying our food.

Unintentional Manipulation of Growth

Humans have also manipulated growth in unintentional ways. Traditionally, biology students are taught to separate selection into two categories— natural selection and artificial selection. This suggests that natural selection occurs without human intervention, whereas artificial selection (also called selective breeding) involves direct human control. In the modern world, however, these categories overlap. Because humans have altered ecosystems on a global scale, we have created massive selective pressures within natural ecosystems. These human-created selective pressures are unintentional but are still reshaping species.

Consider the simple practice of using nets to catch fish and shellfish. On a small scale, netting has little impact on species. But if nets are used throughout the world on a massive scale (as they are today), then they select against large organisms. Only small fish and crustaceans slip through the nets, survive, and reproduce. Thus, over time, we are left with smaller fish, shrimp, lobsters, and other creatures. Nets are nonselective in which species they catch, so they are impacting species that we do not eat, as well as those that we do.

Similarly, hunting for so-called trophy animals serves as a selective pressure. Over time, the trophies become smaller as their larger counterparts are killed. In some cases, the trophy is the whole organism; in other cases it might be horns, tusks, or some other body part. Bighorn sheep, elephants, and other species have already been significantly changed over time due to hunting pressures.

Plants are also affected by unintentional selection. For example, the practice of thinning trees from forests has been practiced for centuries. In some cases, thinning is employed when the resources aren't available to totally deforest an area. In other cases, thinning is well intended when foresters try to harvest trees without destroying an entire forest. In any event, thinning often involves taking only the largest, straightest trees because those are the easiest to haul and process. Thus, over time, the trees that are small and gnarly will have a reproductive advantage.

All of these examples reveal a bitter irony — unintentional selection is having the opposite effect of intentional selection. It is lowering productivity and making it more difficult to harvest the organisms that we most value. Unintentional selection is often bad for humans and natural ecosystems.

Considering all the environments that we've transformed, it is likely that growth has been unintentionally altered in thousands of species. Mowing grass gives a selective advantage to plants with growth regions low on their stems instead of at their tips. Indoor environments give an advantage to bacteria and fungi that grow best at around 72°F. Agricultural fields favor insects and plants that grow and reproduce rapidly so that they are able to evolve rapidly in response to chemical treatments. Roads and highways create novel growth environments beneath them for soil microbes. Fertilizer run-off into waterways favors unicellular organisms that can grow and divide extremely quickly, such as those causing algal blooms. Manipulating the environment not only changes which species are present in an area, it also changes species themselves.

Human Growth

Our manipulation of growth has, of course, included changing ourselves. Over the past few hundred years, people have gotten both taller and heavier. For example, the height of an average European man has increased from around 5 feet in 1800 to around 6 feet today. Not so long ago, many people thought that size differences between some cultures (i.e., Chinese versus American) were due to genetic differences. Although genetics determine much about a person's size and shape, we now know that most of these stereotypes were wrong. From 1950 to 2000, for example, the average height of Chinese children increased about an inch per decade, dispelling a Western myth that the Chinese were innately short. Similarly, Americans in the colonial period towered above people from the Netherlands. Now, however, the Netherlands is the world's tallest country, with the average man standing 6 feet, 1 inch, and women standing 5 feet, 8 inches.

Humans have gotten bigger mainly due to diet. Other factors include exercise, health care, and the make-up of the microbes that each of us happens to be carrying around. Interestingly, by modifying diets and increasing antibiotic use over the past century, we have done to ourselves many of the same things we did to our livestock.

The future of human growth is anybody's guess. Modern sports seem to be a testing ground for growth manipulators of all sorts. Sprinting, swimming, cycling, baseball, football, weight lifting, and other sports have had their moments of controversy as athletes are caught using steroids and other growth enhancers. As world records continue to be broken, one can't help but wonder just how big and fast we can get. NBA basketball players are routinely taller than 7 feet. The tallest person ever was Robert Pershing Wadlow of Alton, Illinois, who was 8 feet, 11 inches tall. Based on other mammals with similar body plans, small modifications to our bodies could make it possible to be even bigger than that. So don't rule out the giants of fairy tales just yet. For that matter, we could also be much smaller. There have been several people under 2 feet tall, such as Gul Mohammed of India, who was 22 inches. So don't rule out elves and dwarves, either!

Only one thing is certain: we are nowhere near the biological and physical limits of what is possible. As our ability to manipulate growth increases in the future, we will have to make decisions about just how far we are willing to push the current biological limits for ourselves and other organisms.

7

WHEN EVOLUTION
TURNS DEADLY

Managing the Evolution of Diseases

Naturally, we're disposed to think about diseases just from our own point of view: what can we do to save ourselves and to kill the microbes? Let's stamp out the scoundrels, and never mind what their motives are! In life in general, though, one has to understand the enemy in order to beat him, and that's especially true in medicine.

— Jared Diamond (Guns, Germs, and Steel, 1997)

"*A* car crash on I-40 kills 2 people." "A suicide bomber in Israel kills 7 people." "Tornadoes in the Midwestern U.S. kill 3 people." "A local man is murdered." Every day, people watch or read the news to find out what is going on in the world. Instead, what we often get is the most trivial information imaginable — "news" that does not accurately represent the most important events of any community, let alone a country or the world.

There's a news industry saying about choosing lead stories: "If it bleeds, it leads." That is literally true – death without blood doesn't usually merit a lead story. Car crashes, bombs, storm deaths, and murders involve blood and appeal to our primitive tendencies. They are like candy for the Neolithic parts of our brain. Never mind that all forms of accidents account for only 6 percent of human deaths. Never mind that war and terrorism account for less than 2 percent of deaths in a typical year.

What would lead the headlines if news accurately represented death? Most commonly we would see stories about heart disease. "Death by microbes" would follow close behind. According to the World Health Organization, infectious and parasitic diseases account for around 20 percent of all human deaths. Thus, if death must lead the news, then once or twice every week the headlines should feature AIDS, malaria, tuberculosis, influenza, diarrhea, or other infectious diseases.

These diseases are especially hideous because they kill people of all ages, including large numbers of children. And they are preventable, raising an obvious question – why don't we pay more attention to them? Why do local news stations in the United States blather so much about murders when far more people die every year from the flu? Why do national and international news media focus so much on terrorism when more people die every minute from infectious disease than from terrorism in an entire year? If we could focus our attention on infectious disease, we could eradicate much of it.

If news represented global reality, a typical lead

story might look something like this:

> "Tonight, malaria continues to ravage much of the developing world. Over a million people have died from malaria this year alone, most of them children under the age of five. Reporting from Botswana is Jane Doe. Hi, Jane." "Good evening, Bob. It has been another heart-wrenching day for families around the world who are dealing with malaria. I've just spent the day with one such family here in Botswana. Several weeks ago, a mosquito bit their healthy young boy as he slept. The family was unaware that he had contracted malaria until three days ago, when he suddenly developed a high fever and chills. As the pain worsened, he began to tremble periodically. His parents took him to a local clinic where he was given medicine, but the fever and chills intensified. His kidneys and brain became severely damaged and he was in terrible pain. I am sad to report that he died today as his parents watched helplessly. As you said, Bob, that's over a million deaths from malaria this year. There have been more than half a billion cases of malaria this year alone! Even when it's not killing, it is keeping people from working, going to school, and leading a normal life. As this family has found, malaria is a devastating problem."

It is difficult to comprehend the human suffering that malaria and other infectious diseases have caused. The very realistic scene described by the imaginary correspondent has been repeated millions of times over the course of history. Likewise, other diseases affect our crops and livestock, contributing to hunger and causing trillions of dollars in economic losses every year. Diseases caused by microbes have been and remain the greatest threat to human societies.

Diseases as Co-Evolving Systems

Many questions about disease that puzzle people share a common answer. Why doesn't last year's flu shot protect us from this year's flu? Why is HIV/AIDS so difficult to treat? Why do diseases become resistant to our medicines? Why do animal diseases jump to humans and vice versa? Why do we need new antibiotics and vaccines? To answer all of these, we may say, "Because all diseases evolve." Last year's flu shot is no longer useful because the influenza virus rapidly evolves as it moves across the world every year. Diseases are dynamic systems that evolve in response to our actions and, in some cases, cause us to evolve in the process.

One way diseases evolve is by forming resistance to the chemicals that we use to kill them. Resistance evolves among viruses and bacteria in response to vaccines and antibiotics. It also evolves in insects and plants in response to pesticides and herbicides, respectively. Just a few individuals of a species may possess genes that give them resistance to a chemical, and after other individuals are killed, these genes are disproportionately passed on to the next generation. Resistance can form in just a few generations, leading to an entire population that is resistant.

It takes anywhere from a few days to a few decades for resistance to form, depending on the organism, the chemical, and the strategy for using the chemical. In general, the faster an organism can reproduce, the faster it will develop resistance.

One of the world's fast-evolving diseases is HIV/AIDS. The HIV virus mutates and evolves so rapidly that a person who contracts one form of HIV may have a great diversity of HIV after several years. The ability to evolve rapidly is one of the reasons AIDS and other retroviruses (i.e., influenza) are so difficult to treat. Try to kill one strain, and you accidentally create several others.

Malaria is another excellent example of resistance and disease evolution. Malaria is caused by several species of unicellular organisms called Plasmodia, which are passed from person to person by Anopheles mosquitoes. In the 1940s, the drug chloroquine was created to fight the Plasmodia, and the chemical DDT was created to kill the mosquitoes (and other insects). Heralded as a miracle cure, chloroquine was effective, safe, and relatively inexpensive. After just a few years, however, some Plasmodia populations evolved resistance to chloroquine. Likewise, mosquito populations evolved resistance to DDT. Resistant strains of the malaria Plasmodia have now evolved in response to nearly all drugs used to treat it, including chloroquine, quinine, sulfadoxine-pyrimethamine, halofantrine, and mefloquine.

Just as malaria and mosquitoes have evolved in response to chemicals, we humans have also evolved in response to malaria. For example, mutations in a hemoglobin gene have given some humans partial resistance to the disease. This trait is commonly found in human populations where malaria has been prevalent, especially in Africa. However, resistance comes at a cost. When a person gets two copies of the resistance gene, one from each parent, the result is a potentially deadly disease called sickle cell anemia. Sickle-cell anemia causes red blood cells to be shaped like the letter C, which in turn causes clumps that stick in blood vessels. Millions of people cope with pain, infection, and organ damage, or even die, due to sickle cell anemia. Indirectly, then, malaria causes suffering even among those who have never been exposed to it.

Because all diseases evolve, our fight with disease is partly a fight to slow down evolution so that our drugs, pesticides, and fungicides remain effective. Management practices can slow down evolution. Unfortunately, though, because people often do not understand the evolutionary consequences of their actions, we have actually done a great deal to speed up disease evolution.

Let's look first at how we've accelerated disease evolution, and then how we can slow it back down.

Speeding Up Disease Evolution

The evolution of diseases probably began speeding up (like much of evolution) around the time that humans invented agriculture. Agriculture allowed much more dense human populations than did hunting and gathering lifestyles. Dense populations are obviously in much closer contact, and they generate large amounts of sewage and garbage. Put these together, and the capacity for rapid disease transmission is greatly increased. In addition, agriculture put humans and other animals in intimate daily contact, allowing diseases the opportunity to hop back and forth between species. Among those that hopped in our direction were measles, tuberculosis, smallpox, flu, and malaria.

These same basic trends continue in the modern world. Populations of people and farm animals continue to grow, meaning that more people share water supplies, create sewage and garbage, and exchange germs. Rapid urbanization is straining sanitation systems; in fact, there are no sanitation systems in some of the fastest growing urban areas of the developing world.

Long ago, a cesspool in one part of the world had no impact on people in another part of the world. But today, modern transportation connects humanity so effectively that diseases can skip around the world in just a few hours. Thus, poverty and disease in one part of the world are everyone's problem.

On one hand, we have made enormous strides against disease over the past 200 years through better sanitation, immunizations, and drugs. On the other hand, we're also driving a rapid evolution of disease that, without better management, threatens to disrupt our progress. Mismanagement (or lack of management) of evolution often speeds up the rate at which diseases evolve. The faster diseases evolve, the more difficult they are to treat.

Perhaps the most obvious way that we're speeding up disease evolution is simply by overusing drugs and other chemicals. The more drugs we use, the faster resistance evolves. For example, we have overused antibiotics on farm animals for decades to increase food yields and lower prices. It is a dangerous trend, because we're reducing the effectiveness of many of the same antibiotics that could otherwise be used for human health. Which would you prefer: cheaper hamburgers or effective antibiotics that can save your life the next time you have an infection?

Meat industries are far from being the only culprits. The amount of pesticides, herbicides, and fungicides that homeowners dump on lawns every year is stunning. Again, the net result is that it speeds up evolution and decreases the effectiveness of the chemicals when they're really needed, which in this case would be for agricultural use. We're essentially trading food security and environmental quality for outdoor suburban carpets. The pesticides, herbicides, and fungicides on lawns are used in addition to fertilizer, lime, seed, aeration, water, and other resources. In all of human history, it is hard to think of anything more resource-intensive and yet so frivolous as the average "well-kept" lawn. We would be far better off thinking of yards as prairies and then treating them as such.

Another way that we speed up the evolution of resistance is by forcing populations of disease-causing organisms through bottlenecks. A bottleneck occurs when a population drops to an extremely low level, and then regrows from just a few survivors. Unfortunately, our use of medicines and chemicals results in bottlenecks much of the time. One chemical is used to drastically reduce a population into a bottleneck, and then only resistant organisms are left. This allows the resistant organisms to reproduce and become a larger percentage of the population. After several generations, a large population with resistance can develop. Thus, diseases can become harder and harder to treat over time.

We see this faulty strategy used throughout society as one antibiotic, pesticide, or herbicide is used over and over until it becomes ineffective. People think, "This chemical worked last time, so we'll keep using it." It seems logical. But the strategy allows resistant populations to grow year after year until the treatment is ineffective. Then we switch to another chemical and use that until it becomes ineffective, and so on. Over time, this gives diseases and pests the opportunity to evolve resistance to all of our treatments. We discuss better strategies in the next section.

Another example is how people often take antibiotics. Should you stop taking antibiotics when you feel better, or after you complete the prescribed regimen? The microbes with partial resistance will be the last ones killed. Thus, if you stop taking antibiotics as soon as you feel better, you may be allowing the resistant ones to reproduce and build up a new, more resistant population. This sort of behavior has played a major role in creating multidrug-resistant strains of many deadly diseases such as tuberculosis and HIV/AIDS. So finish your antibiotics!

Finally, climate change may be one of the greatest contributors to disease evolution over the next century. Put simply, diseases will shift with climate and move into areas where people are not accustomed to dealing with them. Warmer climates have a wide range of brutal diseases that are almost certain to spread. Likewise, disease transmission is likely to increase because of the poverty and migration that climate change is likely to cause. As some areas become drought-stricken or submerged by rising sea levels, millions of people will migrate to other regions. Many migrants will be poor and desperate, and will likely create intense problems of overcrowding and sanitation. Thus, rapid climate change creates a perfect atmosphere for the spread and rapid evolution of disease.

There is only one part of disease systems that we want to evolve quickly — those who get sick. We have been quite successful at breeding and genetically engineering crops and livestock to be resistant to many diseases. In the future, driving our own evolution will also be an important strategy for preventing diseases. We return to gene-based strategies for fighting diseases later in this book.

Slowing Down Disease Evolution

Slowing down disease evolution makes it much easier to treat diseases. Slower evolution means that our drugs will work longer and we have fewer strains of disease to worry about.

An obvious way to slow disease evolution is to use less antibiotics, pesticides, fungicides, and other treatments. This doesn't mean they shouldn't be used when they are needed; it means they shouldn't be used indiscriminately. For example, farmers should minimize their use of antibiotics in livestock and fungicides on crops. The use of chemicals for frivolous things like home lawns should be extremely minimal. This is a classic "less is more" situation—the less we use these chemicals, the more effective they are when used.

In general, chemicals should be used as little as possible when the same results can be achieved using another method, most notably by improving sanitation and hygiene. As an alternative to chemicals, sanitation and hygiene decrease evolutionary pressures in a wide variety of settings. Hand washing reduces the need for human medicines, well-kept fields and healthy soils reduce the need for pesticides and fungicides, and hygienic cattle farms reduce the need for antibiotics.

What is actually happening in much of the world is just the opposite. Chemicals are frequently used as a replacement for hygiene, speeding up disease evolution in the process. A typical industrialized animal farm is again an unfortunate example — only the overuse of antibiotics allows thousands of large animals to be packed close together, standing in their own feces. It is a perfect storm for rapid disease evolution. Slowing evolution means spreading out the animals, using more land, and using fewer antibiotics. That probably means paying more for meat.

Of course, many of us eat far more meat than we need to anyway. All things considered, eating too much meat is one of the most destructive behaviors of modern developed countries. Eating more vegetables and less meat would slow disease evolution, as well as reduce environmental impacts, lower food costs, and in many cases improve health.

When chemicals are needed for killing pathogens, using multiple chemicals instead of overusing just one can slow disease evolution. As mentioned before, using one drug or pesticide over and over inevitably leads to the evolution of resistance. However, using multidrug cocktails or rotating the chemicals used can slow resistance to any one chemical. Sometimes multiple killing strategies can even reduce resistance. For example, chloroquine-resistant malaria in some African countries has become more susceptible to chloroquine after other drugs were used.

Another way to slow evolution is to prevent population bottlenecks from forming. To prevent bottlenecks, we have two options — maintain a larger population of the organisms we normally want to kill or make sure we kill every organism in a given population.

The first option might seem counterintuitive, but it works. If we use chemicals sparingly to reduce population size to a reasonable level (but not so much as to create a bottleneck), then we intentionally allow some nonresistant organisms to survive. Thus, nonresistant organisms keep passing on their genes, making it more likely that the chemical will continue to be effective. For example, it is common practice to plant groupings of nongenetically engineered crops within crops that are genetically engineered with a pesticide. The engineered crop will continue to be mostly effective because whole populations of insects will not become resistant. The trade-off is a slightly lower crop yield, because we have to tolerate insects eating a small portion of crops.

The other strategy for avoiding bottlenecks is to completely wipe out a population of pathogens. In some situations, particularly in human health, this can be an effective strategy. For example, immunization against deadly childhood diseases has been an overwhelming success and saved hundreds of millions of lives.

The risk of trying to wipe out a population is that if you fail, you can end up creating a bottleneck and then a resistant population. Our immunization programs have been lucky in the sense that many deadly childhood pathogens don't evolve very quickly (if they had evolved as rapidly as HIV, immunization would not have worked). Completely killing off a disease is also unrealistic in most field settings such as farms or lawns where the targeted organisms can easily move from place to place.

Finally, we have to be prepared to deal with the

destabilizing impacts of climate change. We have a better chance of sweeping up all the sand from the desert than we do stopping the flow of "climate migrants" into developed countries. Slowing climate change, preparing our health system to deal with different diseases, improving sanitation infrastructure, and ensuring immunizations are all crucial for avoiding future climate-related epidemics.

Scientists are constantly working to create new vaccines, antibiotics, herbicides, fungicides, and pesticides to deal with the constant evolution of diseases and pests. Nevertheless, management practices that slow down disease and pest evolution can allow our chemicals to last longer, saving millions of lives and trillions of dollars each year. There is perhaps no better example of how important evolutionary concepts are to the world.

Evolution Education as a Moral Imperative

We began this chapter by noting that infectious diseases could be greatly reduced if we paid more attention to them. Obviously, that is only part of the story. We have to understand the biology of diseases, as well. Therein lies a problem—in a world where evolving microbes are killing 20 percent of people, a huge proportion of people still don't think that evolution happens. Can you imagine a scientist trying to develop next year's flu shot without understanding how the flu evolves? If half of us don't acknowledge that evolution exists, then half of our collective brainpower is at least partly inaccessible for solving such problems.

Likewise, after nearly eradicating many deadly childhood diseases, many people have forgotten how horrible these diseases are and have stopped getting immunizations. Thus, we are inviting deadly childhood diseases like polio and measles back into societies. Sometimes we can be our own worst enemy.

Perhaps more honesty could get us past this. Teaching evolution and building public health strategies based on evolutionary logic can save millions of lives every year (in addition to the millions it already saves). On the other hand, refusing to teach evolution and refusing to build public health strategies based on evolutionary logic ultimately kills people. If everyone decided to refute evolution, we would be plunged into a Dark Age of plagues and death in as much time as it takes for diseases to evolve resistance to current drugs and pesticides (probably 10–50 years). On the other hand, if we could educate a new generation that embraces evolution, just imagine the progress we could make. In the future, evolution will be deadly only if we remain ignorant of it.

Humans are not doomed to contract deadly diseases. Many deadly diseases have been successfully eradicated in some parts of the world, showing that they can be overcome. It is truly awful that millions still die from diseases that are both preventable and treatable. This is a remarkable opportunity for those wishing to improve the world.

8

GROWING A NEW YOU

Stem Cells and Regenerative Medicine

Objecting even slightly to immortality is a little like arguing against ice cream—eternal life has only been humanity's great dream since the moment we became conscious. And yet we've never had to deal with the possibility that we might actually be able to bring some version of it into being.

—Bill McKibben (*Enough*, 2003)

Consider an amazing fact about yourself: at one point in your existence, you consisted of just one cell. After your father's sperm merged with your mother's egg cell (perhaps not something we'd like to think about), a fertilized egg cell was created. That one cell went on to divide into other cells, which further divided and differentiated into all that you are today. From one cell that could fit on the head of a pin came trillions of cells that can read this book, feel emotions, and contemplate, "I think, therefore I am."

Early embryos are made up of stem cells, which

are unspecialized cells that divide and differentiate into specialized cell types. There are around 200 different cell types in humans, ranging from red blood cells to the intricate networks of neurons. How can just a few cells in an embryo create the diversity of cells that make up a complex, multicellular organism? This "miracle of life" is made possible by stem cells.

Although stem cells make up embryos, they can also be found scattered around adult tissues, where they serve to replace older, damaged cells. For example, our skin totally replaces itself every few weeks, and other organs replace themselves every few years. You are literally "growing a new you" all the time!

Other organisms have stem cells, as well. In fact, the stem cells of many other organisms are more flexible than those of humans. For example, just about any plant cell can be coaxed into producing other cell types and ultimately a whole plant. Likewise, starfish, flatworms, salamanders, newts, and zebrafish can regenerate large portions of their bodies. If they lose an arm or a leg, they grow a new one. This begs the obvious question: "If they can do it, why can't we?" It turns out that humans can replace whole limbs and organs early in development. Perhaps a better question, then, is "How can we extend our regenerative abilities into adulthood like other organisms?" Scientists are steadily finding the answers, unlocking a treasure trove of treatments to cure disease, repair wounds, and extend human lives.

The Potential of Regenerative Medicine

During development, embryonic stem cells produce all human tissues. This has huge implications for treating diseases and injuries. The bottom line is this: because stem cells can generate all tissues throughout the body, they could aid in treating essentially any disease or injury involving tissue damage. Researchers are already exploring how stem cells can be used to treat Parkinson's disease, diabetes, spinal cord injuries, heart disease, vision and hearing loss, and many other conditions.

The potential of stem cells doesn't stop with treating diseases—almost every part of the human body might be replaceable. When people lose a limb or organ today, we replace it with something quite crude. Sometimes we use prosthetic devices (discussed later), and sometimes we use organs from other animals, such as pigs. On other occasions, people receive organs from other people. Stem cells offer the hope that replacement parts will no longer come from foreign sources. If you need a new aorta, it would no longer come from a pig. Instead, your doctor might take some of your stem cells, grow a new aorta, and then put it in place.

Researchers at the Wake Forest Institute for Regenerative Medicine and elsewhere have already grown bladders, two-chamber hearts, and other relatively simple organs in laboratory settings. Dr. Anthony Atala reported the first successful transplants of lab-grown organs in 2006 (bladders given to children and teenagers). This is just the beginning. Hearts? Kidneys? Legs? In time, we may be able to produce any human organ or limb on demand. Imagine a soldier who loses his legs in war, a woman with internal organ damage from a car crash, or a child in need of a kidney. Ghastly injuries, diseases, and disabilities could be healed, and a vast amount of human suffering could be eliminated.

Stem cells used for research and treatment can come from several different sources, including embryos, adult tissues, amniotic fluid, and placenta. Embryonic stem cells are most commonly harvested from one-week old embryos that are grown in a test tube. A common misconception is that these embryonic stem cells are harvested from aborted babies. This is simply untrue. Instead, a sperm and an egg are brought together in a test tube, the egg is fertilized, and the cell begins dividing. After one week, there is a spherical ball of cells called a blastocyst. This is the same procedure used in fertility clinics. Cells from the blastocyst can be taken to generate stem cells. Thus, stem cells for research and treatment can actually come from fertilized eggs in fertility clinics that aren't used and would otherwise be discarded.

The advantage of embryonic stem cells is that they can be coaxed relatively easily into generating other cell types. However, they also come with two disadvantages. Because the cells are genetically different from the recipient patient, rejection is possible. Second, their ease of growth comes with a cost—they can also form tumors.

Another source of stem cells is adult tissues. In our bodies, stem cells routinely heal and regenerate tissue. In a medical setting, adult stem cells have the ability to produce other tissues, as well. When they can regenerate the tissues that are needed, adult stem cells have the advantage of being taken from the patient who needs them so that there is no risk of rejection. However, it is still unclear whether they can be coaxed into acting like embryonic stem cells to produce any other cell or tissue.

Perhaps the most promising sources of stem cells are amniotic fluid and placenta. These cells seem to have the best qualities of embryonic and adult stem cells but without the drawbacks. If a baby's amniotic or placental tissue are preserved at birth, then later in life they may be able to produce other cell types easily without fear of tumor formation or rejection. Stem cells harvested from amniotic fluid and placenta also avoid the ethical issues that some people have with embryonic stem cells.

It may turn out that different sources of stem cells are more appropriate for different medical purposes. It is too early to tell. In the meantime, it seems certain that ethical debates will continue regarding stem cell ethics. Is it unethical to harvest a one week-old blastocyst? Is it unethical not to do so if that leads to the death of full-grown people?

As compelling as these ethical debates can be, there is a longer term issue that has broader implications. If tissues, organs, and limbs can be replaced, then humans could potentially live much longer than we do now. The implications are enormous.

Immortality as Potential Reality

The quest for immortality is probably the oldest example of the human desire to exceed our own biological limitations. In the earliest pieces of literature we know about, such as the 3,000-year-old *Epic of Gilgamesh*, the search for immortality was already underway. In modern times, as Stephen Jay Gould notes, "Ponce De Leon's search for the fountain of youth continues in retirement villas of the sunshine state he discovered. Chinese alchemists once searched for the drug of deathlessness by allying the incorruptibility of flesh with the permanence of gold. How many of us would still make Faust's pact with the devil in exchange for perpetual life?" Take a quick look at the magazine shelves of any grocery store for an answer. The desire to slow and reverse aging has never been stronger.

Throughout much of human history, human life expectancy was between 20 and 30 years. This continued to be normal in most populations through the eighteenth century. During the early 1800s, however, life expectancies began to climb as our understanding of biology and public health improved. Since then, life expectancy has continued to increase to the present day. Today there are a handful of countries (i.e., Japan, Australia, and Canada) with life expectancies over 80 years, and the overall global average is well over 60 years. In summary, people today are living two to four times as long as people once lived. The worst conditions today are equivalent to (or perhaps a little better than) average conditions not so long ago. We have made tremendous progress — without it, most people reading this book would already be dead.

Is there a legitimate reason to think that we could live significantly longer lives than we do now? A pessimist could say that organisms have limited life spans. In humans, for example, there seems to be a biological limit that has kept anyone from living past about 125 years old, regardless of increases in average life expectancy.

A more optimistic view is that some of our cells are essentially immortal. All modern life consists of a long lineage of cells that have been passed from one generation to another for 3 billion years. Your cells came from your parents' cells, which came from your grandparents, and so on for millions of generations. On a cellular level, then, the continuity of life is never broken. Cells can replace themselves indefinitely. We are made out of cells, which offers a proof of concept that immortality is plausible.

We can also point to other organisms that live much longer than we do. Trees around the world have lived thousands of years, including bristlecone pines in California (nearly 5,000 years old) and spruce trees in Sweden (nearly 10,000 years old). Likewise, a variety of animals have lived longer than humans, including some tortoises, whales, clams, and fish.

How might a multicentury lifespan happen for humans? To attempt an answer, we must stray from current science to the realm of speculation and science fiction. We have to dream progress before it can happen, so let's imagine several potential futures. . . As you approach 70 years old, you decide it is time to start life extension treatment. You go to a regenerative medicine specialist, who tells you that the whole process will take several years. The first step is to get your teeth replaced. He numbs your mouth and injects a liquid underneath your old teeth. After a few weeks, your teeth begin to fall out as a new set pushes them out, just as they did when you were a kid. Based on your medical history, the doctor then grows new organs for you—a heart, liver, and kidneys. In one massive surgery, your organs are replaced with fresh ones, sweeping away decades of cholesterol deposits and tissue damage. Once this heals, which is remarkably quick because stem cell treatments help you along, you are ready for a couple of new joints.

Imagine this even better potential future. When you're 50 years old, you go to a regenerative specialist and tell her that you'll be ready for a total-body replacement when you're about 70. She takes a few of your cells and, over the next two decades, grows you a whole new body (minus a brain). When your return, your brain is taken from your present body and placed into the new one. The first couple of years are very difficult as you condition your new nerves and muscles to function. You have to work through treatments and physical therapy for a long time. Eventually, though, you regain control of yourself and start with a youthful body again. You've got a few decades before you have to order a new one.

Or even better, imagine this potential future. Beginning in your thirties, you undergo a regenerative therapy to increase the number of stem cells in your body and heighten your regenerative abilities. The therapy also adjusts genetic activity, turning on some genes and turning off others, so that cells age more slowly. You continue to age, but your body hardly shows it, allowing you to live a quality life for several hundred years.

The future will probably show that several or all of these scenarios are impractical or implausible. But the details aren't the point. The main point is that some method of dramatic life extension will probably succeed using regenerative medicine (accompanied by genetic engineering, which we discuss later). Not only is regenerative medicine likely to increase average lifespan, it is likely to extend the biological limits of human lifespan well beyond 125. It is only a matter of time and research effort. A *lot* of research effort!

In the long term, the final limit to human life spans may be the length of time that brains can function. It may turn out that brains are too complex for us to fully understand, and so life span would forever be limited. However, if it turns out that our brains can be rebuilt and memories restored (by re-creating neuron connections either biologically or digitally), then an individual human could live for thousands of years. The quest for immortality ends with an ironic twist—it comes down to whether brain cells can understand themselves.

Life Spans and Cultural Evolution

Carl Sandburg offered a vision of multicentury living in his poem "Timesweep" (1953):

> I am a three-hundred-year-old galapagos turtle,
> sleeping and eating, eating and sleeping,
> blinking and easy, sleepy-eyed and easy,
> while shakespeare writes a flock of plays,
> while john bunyan sits in jail and writes a book,
> while cromwell, napoleon, lincoln, wilson, lenin,
>   come and go, stride and vanish
> while bryan, morgan, rockefeller, lafolette, algeld,
>   become names spelled and written.
> I sleep, forget, remember, forget again, and ask:
>   What of it?

Don't bother me, brother.
Don't bother a dozing turtle
born to contemplate and yawn.

Would multicentury living lead humans to a sluggish, boorish existence like Sandburg's Galápagos turtle? Maybe longer lives would just give people a longer time to be self-absorbed and indulgent. "There's always tomorrow (*yawn*)." Or instead, would it lead us to greater wisdom, higher purpose, and happiness? Maybe longer lives will give people insights and opportunities that shorter lives could never allow. Quite likely, multicentury living would bring some combination of both bad and good, both boorishness and advancement.

In the long term, regenerative medicine could extend human life spans significantly, forcing us to rethink essentially everything about how civilizations are put together. How will society work if the average person is over 100? Will we overpopulate the planet? Will children be rarer? Will retirement be a thing of the past? Will people become more educated? Allowing ourselves to live for hundreds or thousands of years would lead to radical changes throughout society.

Ironically, increasing life spans could also decrease the rates of some biological and cultural changes. For example, some people might have a greater interest in slowing environmental change and conserving natural ecosystems. Earlier we discussed the idea that environmental baselines shift over time, causing each generation to think that the "normal" environment is whatever was experienced during their childhood. People who live 500 years might better understand the long-term costs and benefits of their actions. They might be less likely to repeat past mistakes.

It is also possible that long-living humans could stifle societal progress. Young people often adapt faster and more willingly to new technologies and new environments. Likewise, new generations sometimes bring with them an improved set of ideas and behaviors that can wash away false preconceptions and biases of previous generations. It is rather harsh to think of death as contributing to progress, yet many prejudices and fallacies might never diminish without death. For example, how might history be different if 150-year-old former slaveholders had been around during the civil rights movement of the 1960s? Sometimes death makes room for positive change.

I've spoken with hundreds of people about this topic. Despite the eloquent quote by Bill McKibben that begins this chapter, many people do not like the idea of dramatically increasing life spans. As with most questions about the future, many people tend to favor today's status quo. They (and perhaps you) think that dramatic increases in life span would be a bad thing. For the sake of argument, let's challenge status quo thinking for a moment by comparing past increases in life span. If we had asked the same question hundreds of years ago, the status quo mindset would have said, "Life spans are better where they are now. Overpopulation could become a problem, and living longer isn't natural." In retrospect, are we better off with a life span of 40 or 80 years? If you can support that 40 is a better average, then it would make sense to project that logic forward to argue that 80 is better than 160. However, if you can support life spans of 80 being better than life spans of 40, then the same logic is likely to apply to another jump from 80 to 160 (assuming that we are adding quality years to life and not just elderly years on the end of life). It is easy to argue for the status quo, in this case that life spans should stay the same, until you realize that past arguments for status quo often look pretty silly.

In any event, life spans do not increase or decrease by public referendum. Instead, it happens as the net result of billions of small decisions. Ask anyone if we should "cure diseases," and they will answer yes. Ask just about anyone if we should do a heart transplant to save someone's life and they will answer yes. Ask anyone if they would like a few more quality years with their loved ones and they will answer yes. Human nature chooses life over death, and in the real world these personal choices overwhelm any big-picture theoretical arguments.

We would be wise to start planning for a future where multicentury living is normal.

9

CLONES CLONES CLONES

Clones

Cloning may be good and it may be bad. Probably it's a bit of both. The question must not be greeted with reflex hysteria but decided quietly, soberly and on its own merits. We need less emotion and more thought.

— Richard Dawkins

*E*very summer, hundreds of human clones come together to meet for a couple of days. What do you think they do? Do they plan diabolical schemes, plot clone wars, and form an evil network for ruining the ethical foundations of humanity? Just imagine being there for the convergence of clones as the evil cackles erupt. Clones are everywhere! Run for your lives!

The name of this get-together is the Twin Days Festival, and its appropriate location is Twinsburg, Ohio. Many of the participants aren't clones because the event is open to all sorts of twins. Nevertheless, there are identical twins galore, each one serving as a living example of a human clone. For the record, when a bunch of clones get together, it is usually for cake and fun.

A clone is simply a genetically identical copy. It really is ironic that clones and cloning have such bad reputation. Just say the word "clone" and you're greeted with a quick, emotional reply such as, "I don't believe in cloning." Images of Darth Vader's storm troopers pop into the minds of many. But say the phrase "identical twins" and you'll be greeted with a smile and genuine fascination. "Where? I want to see them! I wish I had an identical twin!" If scientists working on cloning had called it twinning, they would be a lot more popular at parties.

The bad reputation of cloning is doubly ironic in modern industrial society, since we've already figured out how to make nearly everything around us clone-like. The products of industrial society are identical items—cars, shoes, furniture, houses, phones, and so on. It's hard to think of an object of society that hasn't been manufactured with clone-like precision. Companies do this because it is efficient, making products cheaper. They also do it because consumers demand product consistency. We've been culturally programmed to think that even the slightest product inconsistency is a defect. This mindset burst on the cultural scene during the Industrial Revolution, and since then it has virtually monopolized our food system. Fruits and vegetables must fit a botanical stereotype to make it to store shelves. Any that do not fit the stereotype are sold to a secondary market, where they are hidden in processed foods (which are also expected to be perfectly consistent).

Not only does industrial culture prefer clone-like consistency, it is usually hostile toward anything else.

We are a society that hates clones in theory but demands them in practice. It is an odd and unsustainable hypocrisy.

So should we be fearful of clones? In some cases yes, and in some cases no. As we'll see in a few moments, clones are common in nature. Thus, blanket statements like "cloning is good" or "cloning is bad" turn out to be pretty naive. If you're against all cloning, then you're arguing against identical twins, growth, healing, and bananas (nearly every banana in a grocery store is a clone). That being said, cloning technology does bring both tremendous opportunities and risks. It could both revolutionize and threaten food production. It could create new reproductive options, as well as disrupt ethical frameworks of families. It could increase biodiversity by resurrecting extinct species and decrease biodiversity by creating clonal populations of other organisms. In nearly every case, it could change the course of evolution.

Clones Move from Nature to the Laboratory

Cloning takes place naturally throughout the biological world, from the level of cells all the way to whole organisms. When a cell divides (via mitosis), technically it clones itself. Thus, unicellular life forms reproduce primarily by cloning. Similarly, whenever human cells divide to heal a wound or grow, they clone themselves.

Some animals can clone whole limbs or organs. For example, deer regenerate their antlers each year, and newts can regenerate their arms after they lose one. Finally, whole-organism clones exist in every kingdom of life. Many plants, for example, put out runners along the soil or have sprouts from roots that grow into new plant clones. In echinoderms, some larvae (i.e., from starfish) clone themselves in response to fish predators, presumably to increase the larvae's likelihood of survival by increasing numbers and decreasing size. Likewise, identical twins are common throughout many groups of animals. In summary, clones are a normal part of nature.

As noted earlier, we humans also have clones among us. Identical twins are natural human clones that form when one fertilized egg splits to produce two genetically identical embryos. There are currently somewhere between 25 and 35 million human clones, accounting for around 0.4 percent of the world's population.

Cloning in nature is often a strategic mode of reproduction. From the evolutionary viewpoint of the parent, cloning is advantageous because it passes on all of the parent's DNA. It also allows reproduction to happen in the absence of both sexes, or during emergency periods of environmental stress (i.e., echinoderm larvae). On the other hand, cloning in nature can also be a risky strategy in terms of evolution. Sexual reproduction generates more diversity by mixing the DNA of two organisms in an unpredictable way. This diversity gives populations the ability to adapt more easily over time. If a population relies too heavily on cloning, it can be very susceptible to a disease or environmental stress that can wipe out the whole population.

Cloning entered the domain of human control several thousand years ago, when people began cloning plants as a method of propagation. Just take a piece of a plant such as a leaf or twig and plant it in a suitable growth medium (sometimes normal soil will suffice), and you have created an instant clone. A more sophisticated method is to use tissue culture: in a sterile setting, take tiny pieces of plant tissue and grow them in test tubes or petri dishes with a growth medium. Cloning is frequently used to propagate plants for horticulture, agriculture, and science (and for fun). Examples include strawberries, bananas, orchids, and many others. Although plant cloning has been common for a long time, no one had ever cloned an animal from an adult animal before the 1990s.

In 1996, Ian Wilmut and Keith Campbell broke new scientific ground by creating the first clone from the cell of an adult animal. They took the nucleus from an adult sheep cell and placed it in a sheep's egg cell (whose DNA had been removed). They then coaxed the cell into dividing, implanted it inside a female sheep's uterus, and created a pregnancy. The resulting sheep, named Dolly, had the same DNA as the sheep who had contributed the original DNA. Dolly was an instant international celebrity.

Since that time, scientists have successfully cloned many other animals—wolves, African wild cats, dogs, mules, domestic cats, buffalo, mice, goats, rabbits, horses, gaurs, cows, pigs, rats, ferrets, and others. It has become clear that with trial and error, just about any animal can be cloned in a laboratory.

The techniques have improved considerably since Wilmut and Campbell's first experiments. Nevertheless, laboratory cloning is still inefficient and takes considerable effort to achieve. Furthermore, although many cloned animals have been born and are completely normal and healthy, many have also lived short lives due to respiratory problems and other issues. The techniques are still far from perfect.

So why go through all of this effort? What good does it do to use cloning? Perhaps the most compelling reason is for stem cell research, which often depends on "therapeutic cloning" as a method for harvesting stem cells. Therapeutic cloning is when the DNA of a parent is transferred into an egg cell, the egg is coaxed to divide in a petri dish, and then the cells are harvested after five to seven days. In other words, the cells are never implanted into a mother and a full-grown (or partly grown) clone is never grown. The value of therapeutic cloning is the production of stem cells to be used for healing sick and injured patients.

In contrast to therapeutic cloning, reproductive cloning is when the egg cell with donor DNA *is* implanted into a mother with the intention of growing a full-grown organism. The creation of Dolly the sheep was an example of reproductive cloning.

For industry and agriculture, full-grown clones might produce higher yields and more product consistency in certain instances. For example, a farmer with a corn plant or cow that yields 20 percent more food could greatly increase overall yield by cloning the prolific organism. The same is true for other characteristics, such as taste, efficiency, resistance to a particular disease, and drought resistance.

However, as with clones in nature, this advantage comes at a price—less diversity. Clonal populations can be completely wiped out if a disease or environmental condition strikes that the clones are maladapted to handle. Thus, cloning in industry can be a high-risk, high-reward proposition. For the moment, animal cloning is also expensive and still plagued by health problems, but these are likely to improve as the techniques become more effective and more efficient.

In the scientific community, cloning is also an invaluable tool for research studies of all sorts. Anytime scientists want to investigate the impact of some variable on an organism, using clones eliminates genetic differences between populations in the study. This allows the researcher to know with greater certainty that the observed effects are due to the variable that was experimentally manipulated. Cloned cells, tissues, and organisms have been used to study just about every biological or environmental phenomenon you can imagine.

There are two other more controversial contexts for creating clones that we'll discuss separately – the cloning of humans and the resurrection of extinct species.

Cloning Humans

As we've already seen through the examples of identical twins and other clones in nature, there is no question that clones themselves are ethical. The legitimate ethical questions related to human clones involve the context in which cloning takes place. An example will clarify the point.

Imagine you meet a young man named John, who is an average, everyday guy. It turns out that he is a clone. Before making an ethical judgment about John or his parents, consider these possibilities for how he came to be a clone:

- John has an identical twin brother because of random chance during fertilization.
- He has an identical twin brother because his parents chose to have twins.
- John has an identical twin brother who is 30 years older, who had decided to clone himself.
- He is the tenth generation of cloned Johns.

No one would have an ethical problem with the first scenario, but the ethics get trickier as you move down the list. Note that John himself is exactly the same person in each scenario. Again, the ethical dilemma is not about the existence of clones, but rather the context in which society allows cloning.

Some people argue that the first scenario is the only one that is truly ethical. For these people, randomness is a virtue. They don't think that anyone should be able to make reproductive choices about their offspring other than choosing a sexual partner (and in some cultures, even that choice is prescribed). Others argue that the second scenario is acceptable because functionally it is the same as having identical twins by chance. For these people, there is no virtue in randomness, and choice is generally considered acceptable. Most people are instinctively opposed to the third option. Indeed, it is easy to imagine a few vain people trying to clone themselves a hundred times. On the other hand, there are some subtle examples where the third situation is trickier to evaluate. For example, what if an infertile couple wants to have children of their own through this method? Or what if two carriers of the cystic fibrosis gene don't want to risk having a child with the deadly disease (and don't want to risk an abortion), and would rather raise a clone instead? I've never met someone who would seriously defend the fourth option. (But I know you must be out there somewhere!)

Another possibility is that cloning could be used to create hybrids between a human and another species. It has long been known that two different cells can be fused together when certain chemicals or electrical impulses are present, creating a genetic hybrid of the two. This works using cells of the same organism, different organisms within the same species, and even organisms of different species. Cell fusion has been used most frequently to create agricultural hybrids (i.e., tobacco, tomato, carrot, wheat, and potato). If we chose to do so, it could also be used to create human hybrids, the likes of which the world has never seen. Humans have imagined exotic hybrids for thousands of years—centaurs, sphinxes, mermaids, and the like. In modern times, Hollywood programs us to think of superheroes. But let's get real. Creating Spider-Man is a million times less likely than creating Deformity-Man. The suffering that would have to be endured to create an eventual success would probably be horrifying.

A much more likely possibility is that more subtle hybrids could be created. Instead of combining whole cells, which is inherently unpredictable, human DNA could be carefully engineered and then reinserted back into a human egg cell. Unlike the cell fusion scenario, the genetic engineering option would be likely to produce healthy offspring with the desired characteristics. These traits could still come from other organisms. The possibilities of genetic engineering will be discussed later.

Some of the most interesting applications of

cloning technology are of a different sort altogether. Cloning makes the impossible possible.

Resurrecting Extinct Species

Earlier, we surveyed the massive ecological change that we've caused across the planet. Thousands of species have gone extinct as the direct result of our actions. It was long assumed that extinctions were permanent, undoable scars on the Earth and human history. But cloning makes it possible to bring back species we previously destroyed. Just imagine it—Carolina parakeets, mastodons, moas, and (dare we say it?) even ancient humans could walk the Earth again. The evolutionary implications are enormous—no longer would evolutionary time be rigidly fixed in one direction. Cloning technology allows us to reverse past ecological mistakes caused by our ignorance and recklessness. Of course, we could also make a whole new set of mistakes in the process.

The first "resurrection" of an extinct species has already occurred, albeit only for a few minutes. The Pyrenean ibex (*Capra pyrenaica pyrenaica*), a Spanish mountain goat, went extinct in 2000. The last one was captured temporarily so that tissue samples could be preserved. Later, after it died from a branch falling on its head (talk about bad luck!), its DNA was transferred into the egg cells of a domestic goat. The fertilized eggs were then coaxed to grow into viable embryos before being transferred to Spanish ibex mothers. One pregnancy was successful, and the resulting offspring was a pure clone of the Pyrenean ibex—a proof of concept that species "resurrection" is achievable.

Unfortunately, the process also showed that cloning extinct species is loaded with technical difficulties. The researchers implanted embryos into seventy goats, resulting in only seven pregnancies. Only one pregnancy led to the birth of a live ibex, which then died after a few minutes due to respiratory problems. Cloning a species for the first time is not easy, to be sure.

Woolly mammoths are another interesting example. As discussed earlier, woolly mammoths were abundant in North America and elsewhere in the world when humans first arrived. Very quickly (in evolutionary time), mammoth populations declined. Hunting combined with climate change proved to be a potent combination, and the mammoths were crushed to extinction. We have already sequenced the DNA of a 20,000-year-old woolly mammoth using tissue samples preserved in ice. We could take an elephant egg cell, remove the DNA, insert mammoth DNA, and then implant it in the womb of a female elephant.

For the sake of argument, let's say that scientists attempt this and resurrect woolly mammoths (a likely scenario). Would we put them into zoos for education and research? Would we try to reestablish a mammoth population somewhere? How would they adapt to modern ecosystems? The biggest obstacle to reestablishing the species, as always, will be the unwillingness of some humans to share the Earth with other creatures.

This topic always brings to mind the science fiction thriller *Jurassic Park*. Is it possible to bring back dinosaurs? Cloning a species is only plausible if you know its DNA sequence or have its intact DNA. Although scientists have discovered some surprisingly well-preserved dinosaur tissues, the DNA has been mostly or entirely degraded. Thus, at the current time, we have no way to know the exact DNA sequence of dinosaurs and so they cannot be cloned. But this is far from the last word on the issue. First, there is always the possibility that intact DNA will be discovered. Second, many dinosaur genes still exist in modern dinosaurs, otherwise known as birds. Paleontologist Jack Horner and others have proposed genetically engineering chickens to re-create the probable anatomy and physiology of some dinosaurs. At the expense of an awful lot of chickens, it is probably possible to create a "dinosaur-like" organism in this way. But this method could never produce a verifiable dinosaur that would satisfy a scientist—without the original DNA sequence, we could never be sure we got it right. On the surface, though, I think we'll know a *T. rex* when we see one—they are pretty hard to miss.

There are at least two intellectually sound arguments against bringing back extinct species. The first is unpredictability. For example, what would happen in modern forests of the southeastern United States if the Carolina parakeet were reintroduced? We can't predict anything with certainty, and this should concern us. It should be our duty to do an honest cost-benefit analysis for every species we might clone and then make rational decisions for each. We can imagine a stepwise protocol for researching resurrected species in closed environments and then making more informed judgments about full restoration efforts.

The second argument is that cloning is a money drain—limited resources for conservation should be focused on saving species and habitats that still exist. There may be some validity to this argument, depending on where research and conservation money is obtained. There is no doubt that saving a living species the old-fashioned way is far cheaper than cloning extinct or endangered species. Nevertheless, in reality there is no reason why both cloning and traditional restoration efforts can't be pursued simultaneously because laboratory funding and conservation funding usually come from different sources. In many cases, they will be complementary activities, since restoring some extinct species is essential for restoring some ecosystems to anything resembling their former state. Carolina parakeets and passenger pigeons are excellent examples—these were dominant and widespread species just three centuries ago.

Generally speaking, bringing back species that went extinct recently is far easier to justify than bringing back ancient species. A possible rule of thumb is that if a species went extinct at least in part because of human actions, then we should resurrect them and return them to their natural habitats (or keep them in zoos). That would mean the return of the Carolina parakeet, woolly mammoths, moas, and a host of other wondrous animals. The odds are excellent that we'll see species like woolly mammoths walking the Earth again quite soon.

Resurrecting Ancient Humans

We haven't discussed the most controversial resurrection possibility of all, one that creates unprecedented ethical dilemmas: bringing back ancient humans. It is absolutely possible and perhaps inevitable.

As an example, let's consider Neanderthals. If the word "Neanderthal" brings to mind a dumb, ape-like man-creature, then you should adjust your perspective to fit the facts. Neanderthals are 99.5 percent genetically identical to modern humans, and our last common ancestor was probably around 400,000 years ago (not long in evolutionary time). Humans and Neanderthals crossed paths in Europe 30,000 to 40,000 years ago. Neanderthals had slightly larger brains (be careful who you call "dumb") and an advanced culture for the time period. They were more muscular and had stronger bones, but were slower and less efficient runners. Many anthropologists think that if they were raised in modern human families, they would talk and act like everyone else does. In all likelihood, the only reason they went extinct is because our more recent ancestors moved in and drove them to extinction.

Could we clone a Neanderthal? We have already sequenced their DNA. We could take a modern woman's egg (it may need to be modified), remove the DNA, insert Neanderthal DNA, implant it in a woman's uterus, and let it grow. From a technical standpoint, the most difficult part would be re-creating the Neanderthal DNA (we know the sequence from a bunch of small DNA pieces, but whole chromosomes would have to be assembled). Other than that, it probably wouldn't be any more difficult than cloning other mammals, which we've done many times before. But we can't trivialize how difficult that can be—attempts at cloning in a laboratory usually fail many times before they succeed. The dangers to human mothers could be severe. Just for the sake of argument, though, let's say that a group of women were willing to carry Neanderthal babies. In this scenario, would it be ethical to bring back Neanderthals? Would many failed attempts be worth the eventual return of a human species? Once resurrected, how would Neanderthals be treated? Could they solve problems that we can't or benefit society in some unexpected way? Or would they create problems? Would they want to be one of us?

There is a compelling argument to be made for cloning Neanderthal tissues, if not whole Neanderthals. Because they are so similar to us, yet still different, it is possible that they are immune to certain diseases that plague modern humans. If we created Neanderthal stem cells and could grow Neanderthal hearts, lungs, and other organs, what human conditions might we be able to cure? What might we learn about our own biology by comparison? Could having greater genetic diversity available allow humanity to survive some future plague?

Whether cloning Pyrenean ibex, woolly mammoths, or Neanderthals, we have to acknowledge that the culture and learned behaviors of extinct species cannot be resurrected. We all understand that being raised in a different environment can radically change who we become. In a sense, then, we are really talking about resurrecting "neo-mammoths" and "neo-Neanderthals." They would have to start cultures anew in a radically different world.

But then, given the rapid pace of change in modern times, don't we all?

PART IV
GENETIC CHANGE

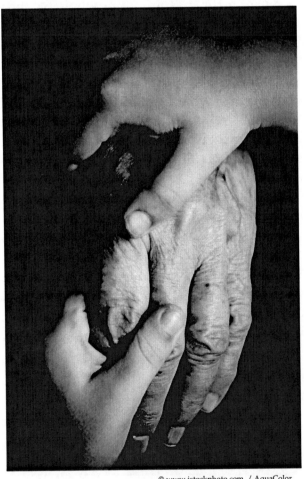

10

WE ARE ALL MUTANTS

Genetic Mutation

We live in a dancing matrix of viruses; they dart, rather like bees, from organism to organism, from plant to insect to mammal to me and back again, and into the sea, tugging along pieces of this genome, strings of genes from that, transplanting grafts of DNA, passing heredity as though at a great party.

— Lewis Thomas (*The Lives of a Cell*, 1974)

*T*here is something eerily beautiful about the music of Béla Bartók. The Hungarian composer was a master at blending pattern and harmony with randomness and discord. Music with a constant pattern eventually gets predictable and boring. Conversely, music that is completely random becomes a cacophony that ceases to be music at all. It is the elegant combination of harmony and discord that creates the effect of Bartók. The dueling forces accentuate one another, bringing the music to life.

Like Bartók's music, DNA contains both pattern

and randomness. DNA is the long string of base pairs—adenine, cytosine, thymine, and guanine (A, C, T, and G)—that passes genetic information from one generation to the next. For the most part, DNA is a very stable molecule. This is important because the stability maintains the integrity of the genetic information. However, DNA is not completely stable. Like Bartók's second violin sonata, DNA has bits of randomness. Frequently, mutations occur and the DNA sequence is changed. In fact, trillions of mutations occur in a human body over a lifetime, 100 to 200 of which are passed to the next generation. To the harmony of life, mutations add a little discord as if to keep things interesting. Bartók would be proud.

It is crucial for life that DNA has these slight imperfections. If DNA never mutated, species would not be able to adapt to changing conditions or evolve over time. A species without mutation would ultimately be doomed to extinction. On the other hand, if DNA mutated too easily, then harmful mutations would build up so quickly that life's processes would unravel. Thankfully, DNA is stable enough to maintain genetic information over time, but is also unstable enough to allow for adaptation and evolution. The pattern and the randomness are both essential for life— they are both part of who we are. We are all humans, and we are all mutants.

Everyone Makes Mistakes

Imagine a DNA molecule as a twisted ladder where every rung represents an A, C, T, or G. This four-letter code holds information that partly determines our looks, personality, and abilities. The segments of DNA that encode traits are called genes. A long strand of DNA, with its genes, is packaged into a chromosome just as a long thread is packaged into a skein of yarn.

A mutation occurs when the sequence of letters is changed somehow. Maybe a C is changed to a T (substitution), or a letter is added (addition), or a letter is deleted (deletion). Maybe a whole section of DNA is replaced, added, or deleted. Maybe sections of DNA rearrange themselves. Maybe a whole chromosome is duplicated too many times, as in Down syndrome, or not enough times, as in Turner syndrome. Perhaps an organism's chromosomes are replicated so that it ends up with several times as many chromosomes, as with plants such as wheat, coffee, bananas, cotton, potatoes, corn, and more. If you can imagine a way that the DNA sequence can get messed up, you can find an example of it actually happening in nature.

"That's interesting, but I'm not a mutant!" you might think. Consider that every time one of your cells divides, it makes a copy of your entire genome (all of your DNA) for the new cell. That is a lot of opportunities to make mistakes. Let's do the math. You have more than 10 trillion cells, each of which contains a complete set of DNA that is about 6 billion letters (base pairs) long (3 billion from each of your parents). That means there are more than 6×10^{22} letters of DNA in a human body. In other words, in the process of growing from a single cell to an adult, we have more than 6×10^{22} opportunities to make mistakes in copying our DNA. Even one mistake in a trillion opportunities would doom us to have lots of mutations. Mistakes are inevitable. Face it, you're a mutant. It is pretty amazing that we don't have more mutations than we do.

How many mutations do we have? It depends on how you think about it. If you want to know how many have accumulated in your body over the course of your life time, that is a pretty hard question to answer. Studies have shown human mutation rates in the neighborhood of 100 to 1,000 mutations per cell by age 15, and perhaps four times that by age 60. This suggests that we all have trillions of mutations! To know a precise range, scientists would have to sequence the entire DNA from many of a person's cells and then compare them to each other. We haven't sequenced enough DNA in individual people to accomplish this yet. Suffice it to say that the total number of mutations that have occurred in your body is a phenomenally large number.

Another way to think about how many mutations we have is in terms of how many are passed along from generation to generation. Only mutations in egg and sperm cells are relevant in this case. The answer, supported by numerous studies, is that about 100 to 200 mutations are typically passed from one generation to the next. Over the 5,000 to 10,000 generations that our species has existed, such mutations have accumulated in our genomes to the point that several million mutations separate any two humans. Even though these differences make up only 0.1 percent of who we are, they are still incredibly important — they account for our genetic diversity and uniqueness as individuals. They also account for thousands of genetic diseases. Thus, as we discuss later, understanding our genetic differences is a key for health care in the future.

The Causes of Mutations

DNA is stable in most normal conditions. In fact, scientists have found intact pieces of DNA that are thousands of years old. Nevertheless, there are a variety of ways that DNA can be changed both naturally and intentionally.

The most obvious way that DNA is changed, mentioned earlier, involves mistakes that occur as it is being copied in the process of cell division. These mistakes alone are enough to make trillions of mutations over the course of one's lifetime.

Another way that DNA mutates in nature is through the normal activity of viruses and bacteria. As Lewis Thomas observes in the epigraph to this chapter, viruses and bacteria routinely pass DNA back and forth. Sometimes this exchange is within a species, and sometimes it is between species. For example, if one species of bacteria in a hospital has resistance to an antibiotic, it can easily pass the resistance gene on to other species. Not so long ago, few would have thought this was possible. The emerging reality is that DNA is exchanged between microbe species all the time.

Microbes are also very clever at exchanging DNA with higher organisms, including humans. The sequencing of human genomes has shown that we are loaded with bits of DNA left behind by viruses and bacteria. Thus, it turns out that the classic idea of an "evolutionary tree" is a little naive because it suggests that the DNA of each species is separated from other species. Now we know that small pieces of the branches of the evolutionary tree can actually hop from one branch to another, sort of like vines.

An example of gene exchange in nature, discovered by researchers at the University of Arizona, involves the color of small insects called pea aphids. Pea aphids can be either red or green. Studies of the aphids' DNA offered a surprise—the red color is caused by carotenoid genes inherited from a fungus. In other words, a fungus passed some of its genes into an aphid, allowing the aphid to turn red. That aphid went on to create new generations of red aphids. In modern pea aphid populations, mutated versions of the carotenoid genes can eliminate the red coloration, resulting (again) in green aphids. Thus, aphid populations can now drift back and forth between red and green to adapt to local environmental conditions.

Radiation is also very effective at mutating DNA. In the 1920s, Hermann Muller discovered that X-rays produce mutations at a rate several hundred times greater than that found in nature. We now understand that many sources of radiation contribute to DNA mutations. The UV rays of the sun are the most obvious, as millions of cases of skin cancer show. Nuclear weapons testing, radioactive waste, and other radiation sources can also increase mutation rates for exposed species.

We have inadvertently altered mutation rates in a variety of other ways. Tobacco use has gotten the most attention, and it does indeed mutate DNA remarkably well. A variety of other substances such as asbestos and dioxins have also been cited for their ability to mutate DNA and cause cancer. A major cause of mutation that gets far less attention is environmental stress. Every time we alter an environment, we alter mutation rates. Environmental stress can take many forms—water or air pollution, disease, drought, hunger, and radiation, for example. Many environmental stresses are closely linked with poverty. Thus, in general, when poverty and stress increase in a human community, mutation rates are also likely to increase. Likewise, when poverty and stress decrease, mutation rates are likely to do the same.

There is an ironic twist here. You might deduce from the preceeding paragraph that areas of poverty will automatically have higher mutation rates. However, that is probably not the case for mutations passed from one generation to the next. Parenthood in poor areas usually begins much earlier than in wealthier areas. Studies show that mutation rates increase dramatically as parents get older. For example, the risk of having a child with Down syndrome increases from 1 in every 1,400 births for a 20-year-old mother to 1 in every 50 births for a 43-year-old mother. By a woman's late forties, mutations are so prevalent that 1 in every 10 births has Down syndrome and over half of pregnancies end in miscarriage. Similarly, studies have shown an increase in mutations in sperm as men age. Thus, mutation rates in the developed world are increasing as people get older and older before having children. The result is more diversity, as well as more disease.

What about mutation rates in other species? From the standpoint of the vast majority of species on the planet, the human rise to dominance has meant that environmental stress has greatly increased. Our pollution, radiation, climate change, and other stressors are driving a wave of increased mutation rates among many species. In some cases, such as with the effects of water pollution on reptiles and amphibians, the effects are well-documented in the form of dramatic anatomical changes (i.e., odd numbers of limbs and multiple heads). We know about these because deformities are easy to spot. Most of the time, however, the effects of mutation are more subtle and thus go undocumented.

Earlier, we explored how evolution happens when natural selection allows some traits to be passed on more frequently than others. We then saw how we are increasing selective pressures and thus speeding up evolution (unfortunately causing a dramatic decline in biodiversity). Now we must add to this equation the fact that humans are also greatly increasing the rate of DNA mutations, which serve to create new diversity. Make no mistake: the rate at which we are destroying diversity far exceeds the rate at which we're creating it. Even so, increased mutation rates raise the interesting point that we are adding even more fuel to the evolutionary fire. Out with the old and in with the new.

The Effects of Mutations

Do not despair, my mutant friends. The good news is that the vast majority of mutations have no effect. Much of our DNA is nonfunctional; only when mutations occur in key places do they cause real problems. In the grand scheme of the world, mutations actually do a great deal of good. They create the diversity between organisms that enriches the world and serve as the raw material for evolution. Mutations often increase the chances that a species will survive over time.

On the level of individuals, however, mutations

sometimes cause great suffering and death. We already know of thousands of genetic diseases in humans that are caused by mutations in particular genes. These include cystic fibrosis, blood disorders, muscular dystrophy, certain types of hearing and vision loss, and many more. In addition, more than 10 percent of pregnancies end in miscarriages due to genetic abnormalities.

On rare occasions, a mutation occurs in a homeotic gene. Homeotic genes are master control switches, telling other genes to do something big, like form a whole limb. Mutating a homeotic gene can lead to wild transformations, such as people with additional arms or legs, flies with four wings, an absence of eyes, or many other radical changes. In addition, some mutations that occur in humans reveal our evolutionary past—webbed feet, tails, excessive hairiness, and so forth.

Although not always referred to as a "genetic" disease, cancer is also caused by DNA mutations. Some genes are responsible for controlling the rate at which cells divide, and when these genes mutate, cell division can spin out of control. When this happens, a tumor begins to form as the out-of-control cells build up. Unfortunately, it isn't so simple as just one mutation causing cancer. For example, a study reported in the journal *Science* found that "The average breast or colon cancer has 93 mutated genes, and at least 11 are thought to be cancer causing." Likewise, dozens of genes are mutated in cancers of the pancreas and other areas. Thus, cancer is caused by a complex array of mutations, greatly complicating the search for cures. To make matters worse, once cells begin dividing uncontrollably, mutations seem to happen more frequently within the cancerous tissue. Mutations beget mutations. The battle to cure cancer is, in fact, a battle against mutation.

If there were a DNA Olympic Games, the gold medalist of the mutation competition would be retroviruses. The genes of retroviruses mutate at an incredibly high rate — a million times faster than human genes. This is a huge problem when retroviruses are making us sick because they change so fast that our treatments can't keep up. It is hardly even a contest. As was discussed in previous chapters, insects can evolve resistance to pesticides and weeds can evolve resistance to herbicides in just a few years. But retroviruses mutate so fast that our drugs can't kill them all within a single person. The most famous retrovirus, HIV, is a classic example. Why haven't we found a cure for HIV/AIDS? Landmark studies in the late 1990s showed that a single strain of HIV mutates in a patient so that, over just a few years, there can be dozens of different variants in the same patient. HIV's ability to rapidly evolve has thwarted our best drugs. In fact, it even thwarts our multidrug cocktails. Treating retroviruses is analogous to trying to draw an outline of your own shadow — you can't draw without the shadow changing. As with cancer and other diseases, HIV/AIDS cannot be cured until we learn to control DNA mutation.

Genetic Testing and Personalized Medicine

Imagining walking into a restaurant. You sit down and a waitress immediately brings over a plate full of food and a drink. "I didn't order this," you say. "But you are here to eat, right?" says the waitress. "Well, yes, but don't I get to choose what I want?" "Let me explain," she says, "You see, we've done studies and we know that this plate of food and this drink are the most effective. More than 90 percent of people like this meal." "But I'm not 90 percent of people! What if I don't like it?" you plead. "Well, then you can order something else after you try the first meal."

We would never tolerate such rigidity in a restaurant, but this is exactly how much of the health care system works. Traditionally, Western-style medicine is based on giving everyone standardized treatments. Nevertheless, it is obvious that everyone is a little different, and in many cases those differences can cause great variability in terms of side effects and drug effectiveness. One person's cure really can be another person's poison. What if we could match particular treatments with the genetics of individual people? Personalized medicine could revolutionize health care by tailoring prevention methods and treatments for each of us.

Personalized health care requires knowledge of each person's unique genetics, and so genetic testing will be a huge part of tomorrow's health care. Over time, knowledge of each person's mutations will be essential for choosing the best treatments just as knowledge of your food preferences is essential for choosing the best meal at a restaurant.

Another part of personalized medicine will be understanding the genetics of microorganisms within a person. Knowing the identity and mutations of the microbes making us sick will allow us to fight them much more intelligently. Likewise, knowing what other microbes are present (but not causing disease) is

134 | Reinventing Life

also useful, because our microbe communities cause subtle changes in our physiology.

In the end, health care outcomes will improve when we treat diseases for what they really are—co-evolving systems that use mutations as fuel for change. The same can be said for managing biological systems in every other area of society.

◆ ◆ ◆

This chapter must end with a confession—Béla Bartók's harmony and discord isn't quite the perfect analogy for the pattern and randomness of DNA. After all, Bartók's music is planned and rehearsed. It seems random at times, but it really isn't. For DNA to be like a symphony, we would have to orchestrate mutation. The tempo of mutations would have to be directed, and decisions would have to be made about which mutations to keep and which to eliminate. We would need to accept our role as composers and conductors of evolution.

Should we direct evolution by altering mutations? Is it our role? To some, it may seem like hubris. To others, it may seem to violate a belief in letting nature run its course. On the other hand, letting nature run its course condemns us to accept genetic diseases, cancer, HIV/AIDS, high rates of miscarriages, and other maladies.

In the end, there is little virtue in standing by while people suffer and die.

11

CROSSING THE SPECIES BOUNDARY

Genetic Engineering

I am not an advocate for frequent changes in laws and constitutions. But laws and institutions must go hand in hand with the progress of the human mind. As that becomes more developed, more enlightened, as new discoveries are made, new truths discovered and manners and opinions change, with the change of circumstances, institutions must advance also to keep pace with the times. We might as well require a man to wear still the coat which fitted him when a boy as civilized society to remain ever under the regimen of their barbarous ancestors.

— Thomas Jefferson
(Letter to Samuel Kercheval, July 12, 1816)

*I*n the last chapter, we learned that genetic mutation is far more common and more natural than most people think.   Although we tend to think of ourselves as genetically stable entities, the truth is that every one of us mutates multiple times every day. Every time one of our cells duplicates itself, a couple of hundred DNA mutations occur. Since the human body has more than 10 trillion cells, that adds up to trillions of mutations, per person, over the course of a human life.   Other species are constantly mutating, too.

Some of our DNA doesn't even derive from humans.   Viruses and bacteria routinely shuttle DNA between organisms in nature, resulting in "human" DNA that is nonhuman in origin. Biologists refer to this as "lateral gene transfer."   Throughout evolutionary history, viruses and bacteria have been shuttling DNA between organisms of every sort.   Most commonly, they deposit their own DNA (which they are also passing readily among themselves).   Lateral gene transfer is a pretty common occurrence in nature, leading to rapid spread of disease resistance genes among microorganisms and other evolutionary events.

Once you realize that DNA is not fixed and is in fact constantly changing, the notion of genetic engineering seems quite innocent.   Changing DNA within an organism and transferring DNA from one species to another are not unprecedented, or even unusual.   Microbes in nature are carrying it out every second.

The only thing unprecedented about genetic engineering is that it transfers control from microorganisms to humans, from randomness to consciousness.   It is pretty difficult to argue that we should give random chance trillions of opportunities to change our DNA, but we shouldn't trust ourselves to do it even once.   Humans have many faults, but we are not dumber or less trustworthy than random chance.

The subject of genetic engineering often sparks an emotional reaction in many people. There is widespread support in some countries for banning or severely restricting genetic engineering. Some activist groups have launched media campaigns and led protests against genetic engineering, advocating "shock and outrage" and denouncing genetic engineering as a "contaminant" and a "dangerous technology." A few groups of more militant demonstrators have gone so far as to vandalize research labs and sabotage experimental field trials.

If these groups wish to be fair, perhaps they should also protest nature itself. Perhaps they should launch a "Campaign to End Cell Division" or a "March to Stop All Microorganisms from Reproducing" since, after all, that is how 99.99 percent of genetic manipulation takes place.

Scientists attempt to view the issues surrounding genetic engineering more objectively. They foresee the technologies greatly benefiting humanity and the environment – as long as we proceed with caution. The Ecological Society of America has stated:

> Genetically engineered organisms have the potential to play a positive role in sustainable agriculture, aquaculture, bioremediation, and environmental management, both in developed and developing countries. However, deliberate or inadvertent releases of genetically engineered organisms into the environment could have negative ecological impacts under some circumstances.

The American Society of Plant Biologists

"believes strongly that, with continued responsible regulation and oversight, genetic engineering will bring many significant health and environmental benefits to the world." The National Academies of Science, which advises the U.S. government, concludes that "an analysis of the U.S. experience with genetically engineered crops show that they offer substantial net environmental and economic benefits compared to conventional crops; however, these benefits have not been universal, some may decline over time, and potential benefits and risks may become more numerous as the technology is applied to more crops." In summary, the science shows that genetic engineering has clear benefits to humanity and the environment, but warns us to proceed with caution.

The simplistic debate about whether genetic engineering is "right" or "wrong" is very unfortunate because it has distracted the public from the truly important questions about the future. How can we use genetic engineering to improve the world? How should the regulatory process be designed to maintain safety while still allowing the timely release of life-saving therapies and improved crops? How can we utilize the benefits of genetic engineering without allowing a small number of corporations to dominate global agriculture? How can we use genetic engineering for humanitarian purposes? How can we use genetic engineering to cure cancer and other diseases? To what extent should genetic engineering be used for human enhancement?

There is no better example of the promise of genetic engineering, and the peril of ignorance about it, than golden rice.

Golden Rice

The World Health Organization estimates that more than 100 million children are not getting enough vitamin A in their normal diets. Every year, vitamin A deficiency causes blindness in half a million children and causes even more to acquire diseases that induce severe diarrhea. In fact, over 1,000 children die from Vitamin A deficiency every day. For two-thirds of the world's population who have the deficiency, rice is the primary source of food.

Dr. Ingo Potrykus, a Swiss geneticist, led a research team to engineer a strain of golden rice that could greatly reduce vitamin A deficiencies around the world. The golden color of the new rice comes from beta-carotene, which is metabolized in the human body to produce vitamin A. People in developed countries typically acquire beta-carotene from fresh vegetables such as carrots. Vitamin A can also be acquired directly from milk, butter, cheese, and liver, all of which are often unavailable to the world's poor.

The project started when Potrykus was brainstorming with Dr. Peter Beyer of the University of Freiburg, Germany, who had isolated genes in daffodils that produce beta-carotene. They quickly realized the humanitarian potential—if the rice grown in poor areas contained vitamin A, then millions of cases of malnutrition could be improved. From 1993 to 1999, the research teams worked to discover how the daffodil genes could be blended with the DNA of rice. After much trial and error, beta-carotene production was finally initiated by blending daffodil, pea, bacterium, and virus DNA and then using a bacterium to carry the genetic instructions into rice.

Despite the immediate impact golden rice would have on world malnutrition and Potrykus's desire to give it away, there were tremendous obstacles in the effort to distribute it. First, he had to free the product from more than seventy patents and legal agreements that involved technologies used in the engineering process.   Then, he had to overcome an effort in Switzerland to ban the export of any genetically modified organisms due to the belief that they are immoral.   Activists did not want a successful genetically engineered food to pave the way for other genetically engineered products. In one instance of public displeasure, several hundred opponents of genetic engineering threatened and taunted Potrykus at a university lecture until he was forced to stop. Eventually the activists successfully lobbied for so many regulations that it became impossible for a noncommercial humanitarian project to succeed.

Potrykus had no choice but to find a corporate partner to help him overcome the regulatory hurdles. The license to golden rice was sold to a company called Greenovation, which then sold it to Zeneca Agrichemicals, which then merged with the agricultural divisions at Novartis, which then became Syngenta, one of the largest agricultural biotechnology companies in the world. The irony of this chain of events is that anti-engineering activists often claim that genetic engineering is just a tool for big corporations, and yet they created a regulatory environment where only big corporations can afford to play. Potrykus brokered a deal allowing Syngenta to sell golden rice in developed countries in exchange for him being able to distribute it for free in poor areas of the world.

Tens of thousands of deaths and millions of

cases of blindness occurred while golden rice was being demonized by activists and caught in one regulatory trap after another. This absurdity led Potrykus to write in an article in the journal *Nature*, "I therefore hold the regulation of genetic engineering responsible for the death and blindness of thousands of children and young mothers." He is probably right. In the eyes of future historians, golden rice may well be a parable about how the protesters' quest for "morality," rooted in an unscientific ideology, ultimately led to great tragedy.

Despite the regulatory difficulties, researchers at practically every major university and research institute are now using genetic engineering to help solve all sorts of global problems. It is no exaggeration to say that a revolution of innovation is taking place.

The Next Generation of Engineered Crops

In the United States, more than 90 percent of soybeans, cotton, corn, and certain other crops are already genetically engineered. The most common genetic modifications involve increased defenses to insects and weeds. For example, "Roundup Ready" crops are immune to the herbicide glyphosate, allowing farmers to spray herbicide and kill weeds without harming the crop itself. Another example is "Bt" technology, which involves crops producing a protein from the bacterium *Bacillus thuringiensis* that is toxic to whatever insects are eating the crop. Although these universal traits will persist, the next generation of genetically engineered crops will include traits for local adaptation, as well.

Many of the best applications for genetically

engineered crops are local in nature—targeted solutions for specific problems. In Hawaii, for example, genetically engineered papaya trees have rescued the whole papaya industry. A ringspot virus was destroying Hawaii's papayas in the 1980s and 1990s. Researchers from Cornell University engineered a resistant tree that was then cross-bred with other varieties. Now more than 80 percent of Hawaiian papaya trees have the genetically engineered trait for virus resistance. Similarly, in China, engineered cotton that is resistant to the destructive cotton bollworm has been widely planted. Analysis has shown that using the engineered cotton even controls the bollworm on nearby nonengineered plants.

As mentioned previously, rice is a staple crop throughout the world, especially in poor areas. Researchers at University of California, Riverside, and the International Rice Research Institute have created varieties of rice that can withstand being submerged under water for almost two weeks, which can save crops during years of flooding. At UC-Davis, rice has been engineered to have greater salt tolerance. Others are working on rice that will be more resistant to drought, cold, iron toxicity, and other stresses. Added to the nutritional benefits of golden rice, these new traits could have enormous humanitarian benefits.

Colorado State University researchers have created plants that can change color when certain pollutants or even explosives are nearby. This could allow the plants to serve as a warning system during a terrorist attack, industrial accident, or when landmines are left behind after wars.

Partial solutions to the world's energy needs are also being addressed by genetic engineering. At many universities, organisms are being engineered with improved characteristics for producing biofuels (i.e., tolerance to glucose and ethanol). Plants, bacteria, yeast, algae, and other organisms have been engineered for this purpose.

In Australia, field trials have been promising for insect-resistant cotton, drought-tolerant and salt-tolerant wheat, and barley that is boron-tolerant and fiber-enriched. Researchers have also created bananas that are fungus-resistant and fortified with vitamin A and iron.

In South Africa, researchers have engineered corn that is resistant to the maize streak virus. The virus, which is endemic to Africa, can completely destroy a farmer's crop in a bad year. Because corn accounts for more than 50 percent of calories consumed in some African regions, the new corn could help the continent to become more stable and self-sufficient.

We could go on and on with examples of genetic engineering being used to solve specific problems and improve particular crops. Basically, if you can imagine it, then several research labs are working on it. Genetic engineering will allow crops of the future to be better tasting, more nutritious, more tolerant of environmental stresses, and less allergenic. Foods will also last longer before spoiling, allowing food to be distributed more easily. Finally, and perhaps most important, there will be much more food grown per acre, meaning that we will need less land to grow crops. This creates the opportunity for millions of acres to be wilderness instead of farmland.

A common criticism of engineered crops is that they allow a small number of large corporations to control an agricultural system. If every farmer is using

Roundup Ready, Bt, or a few other corporate systems, then there will be less agricultural diversity, more corporate control, and little economic benefit for the farmers themselves.

In many situations, this criticism has been at least partly true. Nevertheless, this is a problem caused by patenting and regulatory systems, not genetic engineering itself. As long as the regulatory environment is biased against engineered crops, large companies will continue to dominate because most smaller players can't afford to get products approved. As with golden rice, many of the best uses of genetic engineering, especially those with benefits for poor and developing nations, are having trouble moving from the laboratory to the field.

Despite its potential, genetic engineering is not a panacea for agriculture and food supply. Agriculture is taking a serious toll on the planet. Global population is growing, soils are being degraded, and water supplies are being depleted. Perhaps most important is that climate is changing, making traditional agricultural methods obsolete in many regions. Genetic engineering can help remedy all of these problems, but it cannot be a complete solution by itself. We also need to embrace sustainable practices that build soils, reduce unnecessary herbicides and pesticides, increase biodiversity, reduce water usage, and distribute food more efficiently.

In the future, if we are wise, we will stop pitting different agricultural systems against one another. For example, both modern scientific farming and traditional indigenous agricultural systems both have their place. In a world with rapidly changing environments and cultures, we need the tools and techniques of every agricultural system at our disposal to help individual regions cope with their own unique circumstances. Clinging to traditionalism will not work when tomorrow's climate is different from yesterday's. Likewise, using a small handful of corporate methods all over the planet is unlikely to benefit such a wide diversity of peoples and environments.

A sustainable and equitable future looks like this: crops and livestock are genetically engineered with specific regions and peoples in mind so that the local cultures are empowered and crop biodiversity is maintained. It is a world where ordinary people control their technology instead of technology controlling them. This is the best future we can hope to attain.

Genetic Engineering for Human Health

Genetic engineering is also enormously useful for improving human health. It can be used on other organisms to produce drugs and directly on humans to reverse harmful mutations.

The first drug produced by genetic engineering was insulin, approved by the FDA in 1982. Before then, people with insulin-dependent diabetes had to inject themselves with insulin produced from cows or pigs. Although effective, cow and pig insulin increased the chances of allergic reactions. The company Genentech genetically engineered the bacteria E. coli so that it would produce a human version of insulin. Since this first success, therapies have been engineered to treat multiple sclerosis, strokes, dwarfism, cancer, and a wide range of other diseases. By moving medicine away from using chemicals and parts derived from other animals and cadavers, genetically engineered products have resulted in higher success rates and fewer allergic reactions.

Genetic engineering is also invaluable as a method of disease prevention. Human papillomavirus (HPV) causes genital warts and is also the main cause of cervical cancer, which kills thousands of women each year. The pharmaceutical company Merck produces a widely-used vaccine for HPV. When first released, the vaccine represented a medical breakthrough for being among the first products to actually prevent a form of cancer. It is little known by the public that the vaccine is produced using genetically engineered yeast (that is, in fact, the only way it could ever have been produced).

All around the world, companies are developing new genetically engineered drugs to fight cancer. Some will prevent various forms of cancer, and others help keep cancer in check. For example, injecting tumor suppressor genes can slow some tumors.

Genetic engineering can also be used to fix genetic diseases, birth defects, and a broad range of other harmful mutations. Most people don't like the thought of "engineering" humans because we don't like to think of ourselves as simply a mechanical product. We prefer to feel "cured" instead of "fixed." With this in mind, medical practitioners have adopted the gentler phrase "gene therapy," which includes a broad array of methods for using genetics to treat disease. Whatever we call them, genetic approaches can save lives and restore health.

For example, researchers at the University of Pennsylvania, Children's Hospital of Philadelphia, and

University College London have corrected a gene defect in the eyes of people born with severe blindness and partly restored their sight. Only one injection was needed to produce the dramatic results.

In another case, researchers at the National Institutes of Health genetically engineered the lymphocytes of cancer patients so that their cells would recognize and destroy cancerous cells. Several patients with rapidly advancing and deadly forms of cancer were cancer-free a few months after the gene therapy. Yet another example is the treatment of HIV/AIDS. Many research groups are giving HIV-infected patients new genes to help fight HIV by removing blood or bone marrow, introducing new genes to immune cells, and then re-infusing the cells back into patients.

As we know from the last chapter, several thousand genetic diseases are caused by DNA mutations, and a variety of infectious diseases use mutations of their own to outsmart our drugs. Genetic engineering has the potential to prevent, manage, or cure the majority of genetic diseases, and also to develop more effective, adaptable cures to infectious diseases. Thus, genetic engineering is sure to become more and more important as a health care tool.

In the process, it is also quite likely that genetically engineering ourselves will become common.

Extreme Forms of Genetic Engineering

It would be great fun to hop into a time machine and go back to witness the very first time that someone extracted and drank milk from a cow or goat. It may have seemed unthinkably disgusting to people at the time. Was it a caveman dare? Or a tribe who was desperately hungry? Or an early experimentalist?

Whatever the case, it goes to show that what seems extreme and unnatural to one generation or culture can be totally ordinary to the next. We eat and drink other organisms, even some that were once considered poisonous — tomatoes, for example. Likewise, ancient human cultures would have been shocked to hear that we now replace our organs with those from other animals, or graft plant species together to make them grow as we like.

What are the more "extreme" genetic engineering projects of today that may seem ordinary tomorrow? We'll start with the more mundane and work our way to the more radical.

Although genetically engineered animals are not yet a major part of agriculture, they are coming in a big way. They will be much more efficient — growing faster, requiring less food, and producing less waste. Eventually they will also produce leaner, lower-fat meat.

If we choose, we could also grow meat in an industry setting that isn't really from an animal, per se. The same cells that divide and grow to produce "meat" in an animal can be coaxed into growing "synthetic meat" in a laboratory. Some would argue that current meat production practices have become so miserable and unethical for animals that synthetic meat would be an ethical improvement. Public opinion will ultimately decide.

In the realm of humans, genetic engineering has huge implications outside of just fixing health problems. We will almost certainly see human enhancements of all sorts. Among the first could be people engineered to be slimmer and more muscular, both of which have already been done in mice and monkeys. Eyesight could be greatly improved, perhaps even allowing us to see wavelengths of light that are currently invisible to us—maybe to the point where we could lessen our need for lighting and electricity use. Intelligence is more complicated and more impacted by one's environment, but it, too, could be genetically enhanced.

Undoubtedly many people will protest vehemently at the notion of human enhancement, and it will probably be banned in some places. At the same time, though, the competitive pressures to use it will be enormous. Will the landscape of global power shift based on who embraces genetic technology and who doesn't? It is possible that we are in for a sort of genetic arms race or, if you view it more positively, a global revolution in genetic innovation. No matter what you call it, it would be an evolutionary sprint.

Genetic engineering will also alter future athletic events. Unlike the use of performance-enhancing drugs, which a medical exam can detect, it will be nearly impossible to prove many cases of genetic enhancement. In fact, some professional sports may be forced to adopt something like the NASCAR model, in which every competitor has access to essentially the same technology. That way, at least you create a level playing field. Genetic enhancement isn't restricted to professional sports—we have probably already seen the last Olympic Games without genetically engineered performance.

Some argue that genetic enhancement could ruin sports, and in some cases they may be right. On the other hand, was it ever really fair that a few people are lucky enough to be able to hit a baseball 450 feet, run 100 meters in under 10 seconds, or jump from the free-throw line and dunk a basketball? Sports are already dominated by genetics more than we would like to admit. The real decision is whether we prefer sports to be driven by genetic chance or genetic design.

The advanced research divisions of militaries around the world are applying genetic engineering to create unprecedented weapons and more capable soldiers. If they succeed, future war zones might feature troops with superhuman metabolisms, attack bees that follow orders, genetic programming with remote controls, rapid-healing stem cells, or living machinery.

Other more benevolent uses of genetic engineering are also feasible. One of the more thought-provoking ideas is the "open-source organism" concept. Open-source projects rely on the good will of the community to collectively design something. One designer makes a change, another improves it, another improves it some more, and so forth. When an open-source project becomes popular, it is hard to beat because everyone together knows more than a small group. Wikipedia is a widely popular example.

Now imagine if the global community designed organisms collectively, just like it now collectively maintains Wikipedia. One group contributes genetic information for superior sugar metabolism, another adds DNA repair mechanisms, and so forth.

Some researchers, such as the BIOFAB group, are developing free standard DNA parts that could be used to create designer microbes. Eventually, it might be easy enough for novices to participate. You could choose the traits you want instead of having to understand exactly how the genetics works, much as we buy a television or computer because of what they do even though we don't usually build one ourselves.

The open-source concept is revolutionary not only for the novel organisms it might produce, but also for the evolutionary process that it creates. Evolution would have expanded from a process of natural selection to include artificial selection, then independent design, and finally community design. Evolution could actually become a democratic process.

Finally, we can imagine more large-scale genetic shuffling between higher organisms. Mythologies and religions are full of fantastical creatures — centaurs (part human, part horse), chimeras (mix of lion, goat, and snake), sphinxes (mix of woman, lion, and bird), angels (human-like creatures with wings), and so on. Modern superheroes also embody imagined human-animal hybrids — Spider-Man, Batman, Wolverine, and so forth. Obviously, many of these are ridiculous from the standpoint of what is possible. But some of their characteristics are absolutely achievable. It is our ethics, and not our science, that would keep some of these things from happening.

Leaving Our Primitive Times Behind

Once upon a time, our ancient ancestors scratched lives from nature. When food presented itself, they ate. When it did not, they starved. In the real world, "leaving nature alone" really means subjecting ourselves and our families to merciless and random suffering. Eventually humans took more control of animals and plants through agriculture, and then civilization took off. Today we can hardly imagine how harsh the preagricultural existence must have been.

Fast forward to the future. Our descendants may look back at us in the same way we look back at our ancestors. They will briefly consider what it was like for genetics to be random and uncontrolled, but they won't really understand. They will see us as poor wretches who struggled to do the best we could under tough genetic circumstances. Just imagine your great-great-great-great-grandchild strolling down a museum hallway, looking at artifacts of the "pre–genetic engineering era" and wondering aloud, "What would it have been like to live during such a primitive time?" Just as today, some of our descendants may not even believe that they evolved from us. "We couldn't have come from those monkeys," they may say.

12

HUMANS BY DESIGN

Exploring Our Reproductive Options

> I'm advancing a theory that we were not
> biologically designed to remain in our present
> stage any more than a tadpole is designed to
> remain a tadpole.
>
> — William Burroughs
> ("The Four Horsemen of the Apocalypse,"
> Lecture to the Eco-Technic Institute, 1980)

I am writing these words as I sit in a doctor's office with my wife. She is nine months pregnant, and any day (or any moment) we expect the birth of our second child. Anyone who witnesses the development and birth of a child can't help but be amazed by the whole process, and amazed by life itself. It leads you to contemplate the future. What kind of world lies ahead for our children? And what kind of children lie ahead for our world?

In this particular doctor's office, we are surrounded by magazines with titles like *Conceive,*

Bundle, and *American Baby*, as well as informational pamphlets on childbirth and maternal health. It reflects a simple, and traditional, view of childbirth. Conceiving and having a baby can indeed be a biologically straightforward affair, and historically it has been.

In the modern world, however, there are myriad reproductive options, many of which are relatively new from a historical perspective. Thirty-year-olds can reproduce in a very different way than their grandparents did. An even greater array of options will soon be possible, and our descendants a few decades from now may choose to reproduce in a completely different way from what we choose. "Family planning" has never changed so quickly or been so complicated.

Interestingly, there are no educational materials in our doctor's office to explain the full range of reproductive options. There is nothing about in vitro fertilization, artificial insemination, fertility drugs, abortion, and other options. And there is certainly nothing about trait selection, gene therapy, or other genetic technologies. On one hand, you can understand why—many people on the traditional path of childbirth don't want to be bothered with these options. Others would be offended by their presence in a waiting room, and some would be disappointed to find that some options aren't widely available or are too expensive. On the other hand, they *are* options, and people have a right to know.

This chapter lays out our reproductive options for the future in a completely honest and unfiltered manner. It is not intended as a medical guide for individual families. Instead, it illustrates the spectrum of possibilities for the human species. If we look at the big picture, how can humans reproduce? How might our collective decisions influence the future of the human species?

A Simplified View of Reproductive Options

The diagram below shows a simplified view of modern reproductive options. Starting at the top and working your way straight down the middle represents the current conventional means for reproducing. If you want a baby, you have sex, get pregnant, go to the doctor and get tests done to ensure the baby's health, and then give birth to the child. This is a simple path that would satisfy most couples.

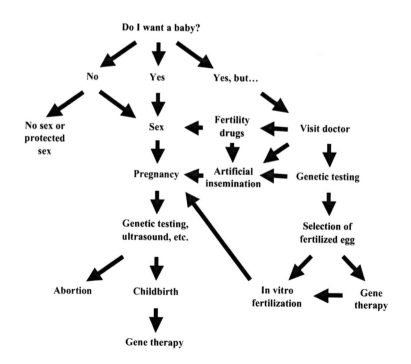

If only the real world were so simple. A huge number of complications can arise, and often do. Look at the diagram again, and this time go down the right side through the "Yes, but..." option. A common "Yes, but..." scenario is "Yes, I want a baby, but we have infertility problems." Infertility rates among U.S. women are around 11 percent at age fifteen and increase to 27 percent by age forty-four. Likewise, male sperm counts decline significantly with age. For all ages, there appears to be a global trend of increased infertility for both men and women, although the underlying causes for this are debated. Infertility is driving more and more people to doctors for help.

Other "Yes, but..." scenarios abound. "Yes, but I've had multiple miscarriages." "Yes, but I've had an abortion before." "Yes, but I'm over forty years old." "Yes, but I am a carrier for a deadly genetic disease." "Yes, but my partner and I are gay." "Yes, but I don't have a partner." There are also many "Yes, but..." possibilities that involve parents wanting more control over the reproductive process. "Yes, but I'd like twins." "Yes, but I'd like a girl since I've already had a boy." "Yes, but I'd like to ensure that the baby has my brown hair and my spouse's eyes." "Yes, but I'd like my baby to have a set of genes that give him or her a better chance of playing professional sports." "Yes, but I'd like my baby to have an excellent memory." Thus, people's reasons for wanting more reproductive control range from medical necessity to desire for enhancement.

Whatever the "Yes, but...," the next step is to go to the appropriate doctor. For issues related to infertility, miscarriages, and other common problems, finding an appropriate clinic isn't too hard. For example, more than 12 percent of U.S. women between the ages of fifteen and forty-four have used infertility medical services. Currently, however, finding a clinic that enables particular types of trait selection is much trickier. Although there is a doctor or geneticist somewhere in the world who can at least partly address every scenario above, finding them may not be easy. This may be a temporary situation that will change as reproductive technologies become more widespread. On the other hand, if society rejects certain technologies, they may remain on the fringe of medical practice.

Once you find the right doctor, if your issue is conception or infertility, then fertility drugs, artificial insemination, or *in vitro* fertilization might be the solution. If your issue involves the traits of the offspring, you'll move down the lower right of the diagram. Genetic testing (in this context known as "preimplantation genetic diagnosis," or PGD) would tell you which eggs, sperm, or fertilized eggs have the traits you want or don't want, at which point you can make choices. In some cases, especially when the desired traits are not in the egg or sperm, gene therapy could be used to introduce the desired genes. Finally, *in vitro* fertilization allows the chosen fertilized eggs to be inserted into a woman's body to initiate a pregnancy.

Although choosing particular genes for a child is pretty straightforward, choosing physical traits is much more complicated. It is important to note that many traits result from complex combinations of genes interacting with environmental cues. Thus, our ability to choose traits is clearly limited. That being said, our ability to choose traits is also just getting started. Every decade of the twenty-first century will probably usher in a new wave of traits that can be selected.

We are skipping lots of details and other possibilities, of course. Future technologies will probably change the flow chart. Nevertheless, this simplification should allow us to address our central questions—what sorts of choices can humans make, and what does this mean for the future of humans?

Possibilities for Human Trait Selection

Just as humans have made reproductive choices that have driven the evolution of pets, livestock, and crops, we make choices that drive our own evolution. Of particular importance for our future evolution is parents selecting traits for their children.

We can think of selecting the traits of our children on at least five different levels. With each successive step, our options for trait selection increase by an order of magnitude. Likewise, with each step, the capacity for rapid human evolution increases dramatically.

The first level is simply choosing a sexual partner. We don't usually think of mate selection as a method of selecting our offspring's traits, of course. We want someone that we find attractive, intelligent, fulfilling, self-supporting, and so forth, and usually we choose based on how happy a person makes us. But let's be honest—choosing a sexual partner is also choosing a family. We want someone who is attractive and pleasing, but these same traits will be passed on to children, allowing them to successfully reproduce and give us grandchildren. Choosing a sexual partner is effectively a choice about one's descendants.

The second level of trait selection is choosing the genes that are passed on within the genomes of two people. For example, my son's DNA is half from me and half from my wife. But that means that only half of my DNA was passed on, and the same for my wife. What genes did and did not get passed on? Did the randomness of blending genes together accidentally create a genetic disease, or perhaps a remarkable trait of some sort? It is already possible, using current technology, to make many decisions about the genes that are passed on within a family. This has been going on in clinics for several decades. The more popular it becomes, the faster humans could evolve because certain types of decisions will be popular or even universal (everyone might select against childhood cancer, for example).

The third level of trait selection begins to take us into the realm of the future—choosing the genes for a child using the full array of genes from across humanity. In this scenario, people could choose human genes that aren't present in their genome or in their partner's genome and use gene therapy to insert them into a fertilized egg. Thus, choosing a sexual partner would no longer be the limiting factor for an offspring's traits. By choosing genes from across humanity, a broad-based wave of selection could ensue that would favor certain traits and shun others. This would represent a landmark change in human evolution—for the first time all human genes could undergo a *conscious* selection process.

The fourth level may result in the most controversy and the most protest: choosing genes for children from across the entire spectrum of living things. This does not mean creating human-animal monstrosities, but using a small set of genes that happen to be advantageous for specific purposes. For example, Rhesus monkeys have a gene that gives them resistance against the HIV virus. Certain genes from other organisms might give us the ability to heal faster, live longer, see better, or increase our abilities in other ways. The obvious arguments against using nonhuman genes are unpredictability and ethical concerns. But when transformative enhancements are within our grasp, it seems nearly certain that someone will attempt them. To go further, since most people have no problems eating the flesh of other animals or accepting animal parts for organ transplants, it seems likely that people might one day accept animal genes to enhance themselves. As Emily Dickinson reflects,

> The pedigree of honey
> Does not concern the bee;
> A clover, any time, to him
> Is aristocracy

The final frontier of human trait selection would be the adventurous option of choosing traits from anywhere they become available. This includes the previously mentioned genes from humans and other organisms, as well as genes that are synthetically created in laboratories. Whatever traits could be invented by scientists could be introduced into our genomes. The new field of synthetic biology has already led to the creation of novel microorganisms and molecular machines. On one hand, this could be transformative for human evolution, as it allows us to transcend the traditional evolution process. It could lead to humans with traits that are totally unprecedented in current living things. Practically speaking, however, it might be difficult (at least at first) to engineer genes that are better than what we could find in nature.

Do we want conscious human thought or unconscious evolutionary logic to determine our fate as a species? Erwin Schrödinger asserts, "Concerning the question of whether biological development is still to be expected in man. . . . We must not wait for things to come, believing that they are decided by irrescindable destiny. If we want it, we must do something about it. If not, not." If we choose unconscious logic, or we choose nothing at all, then we are willingly accepting a cruel future. After all, suffering and death are of no concern to unconscious evolutionary forces. On the other hand, if we choose conscious thought to determine our evolutionary future, we will have adopted a tremendous responsibility for which we may or may not be prepared. Both conscious and unconscious routes come with their own risks.

The Risk of Genetic Homogeny

A popular concern about trait selection is the possibility that too many people will choose the same traits and human diversity will be decreased. Would we all end up looking and being the same, or at least more the same? It does seem probable that certain disease-related genes would decline, and some genes conferring enhanced abilities would increase. Beyond these generalizations, it is difficult to predict what individuals would choose.

Would choices be made the way that people choose clothing, in which case it is embarrassing to cross paths with someone who is wearing the same thing? Or would choices be more like the way people choose music, where certain styles are popular for a time and then become passé? Or would choices be like fast food preferences, where common decisions result in a few restaurants dominating the food system and reducing overall diversity? Or would choices be like how people choose art, where there is great diversity in terms of what people like and dislike, and individual preferences are often based on cultural factors? We really do not know what people will choose and how choices will change over time. Ironically, making choices about our own evolution actually makes evolution more difficult to predict.

The choices that people make about trait selection are ultimately based on their values and their view of human nature. Are humans supposed to suffer and deal with the circumstances that nature presents to us? Or are we supposed to take control of our own destiny and try to improve our condition? What is the goal of humanity? Perhaps the most promising guarantee of diverse genetic decisions is that there will never be consensus on these questions. As Nietzsche reflects, "A thousand goals have there been so far, for there have been a thousand peoples. . . . Humanity has no goal." It seems completely against human nature for all people to have one opinion or to pursue one single common goal for any substantial length of time. For this reason, the fears of genetic homogeny among humans may be unfounded in any situation where individuals make their own choices, assuming their options are plentiful.

The most dangerous possibility for trait selection is that governments will force people to make certain decisions or will severely limit choices to make certain decisions overly common. We can imagine scenarios where authoritarian regimes force a reproductive scheme on a people based on prejudicial ideas about what is "better" for humanity. We can also imagine well-intended people wanting to construct a "level playing field" for everyone in society, and thus arguing for a limited set of genetic options for trait selection. Finally, we can imagine governments creating so many regulations to oversee trait selection that only a few large corporations can afford to provide genetic options, inevitably leading to very few choices. All three cases could be disastrous for human diversity. The surest way to destroy genetic diversity is for governments to limit individual choices and/or for a small number of corporations to dominate a human gene market.

Protecting the ability of individuals to make their own reproductive decisions is the surest way to ensure genetic diversity.

Transcending the Human Condition?

There are at least four possibilities for future human evolution. First, we could stay essentially the same for a long period of time. Although this might sound appealing, this is probably the least likely possibility. Only a small fraction of species lasted unchanged for long periods of evolutionary history. The second possibility is that we will go extinct, ending our branch of the evolutionary tree. Most species throughout history have suffered this fate, and it could certainly happen to us. Third, we could change over time until we eventually become another species. Finally, we could diversify and branch into multiple species. It is easy to imagine scenarios in which any of these could become future reality.

If we move forward with trait selection and other reproductive technologies, is it possible that we could rapidly evolve into another species and begin to transcend the "human condition"? Yes, but probably not any time soon.

Most people have unjustifiably narrow views of humanity. Consider a few "oddities" of natural human variation that blow away our stereotypes about what it means to be human. Full-grown humans have ranged from under two feet tall to almost nine feet tall, a remarkable difference of four times! Francesco Lentini of Sicily had three functional legs. Ching Foo of Shensi had a permanently blue face. Weng, "The Human Unicorn" from Manchuria, had a thirteen inch horn growing from the back of his head. Thomas Wedders of England had a nose that was more than seven inches long. Chelsea Keysaw of Wyoming kept growing teeth even after her permanent teeth had come in, much like a shark keeps replacing teeth throughout its life. Joe Laurello, "The Human Owl," could twist his head 180 degrees so that he was "facing" his rear. Johnny Eck was born with no lower half of his body, yet was still an accomplished performer and painter. Numerous people have been born with tails, thick hair on their entire body, or webbed fingers and toes. Others have lived as a "cyclops" (only one eye), conjoined twins (two people fused together), or without the ability to feel pain. We could go on and on with examples.

Given natural human diversity, we must define "humanity" and the "human condition" quite broadly. If we define it broadly, then it seems unlikely that human nature, itself, is under any immediate threat. Of course, that could change in the future. At the moment, though, it is really more what we think of as an "average" human that could change quickly.

Currently, most people support using reproductive technologies to change individual humans in small ways that would improve health and wellness. At the extremes are some who are repulsed at trying to change humanity as well as others who embrace "transhumanism"—future humans with enhanced capabilities. Julian Huxley, who popularized the framework for thinking about evolution that continues to this day, points out that human evolution is still ongoing. He coined the word *transhumanism*, and argues that we should embrace self-improvement:

> The human species can, if it wishes, transcend itself—not just sporadically, an individual here in one way, an individual there in another way, but in its entirety, as humanity. We need a name for this new belief. Perhaps *transhumanism* will serve: man remaining man, but transcending itself, by realizing new possibilities of and for his human nature.
>
> "I believe in transhumanism": once there are enough people who can truly say that, the human species will be on the threshold of a new kind of existence, as different from ours as is from that of Pekin man. It will at last be consciously fulfilling its real destiny.

Humanity remaining human, but transcending itself. . . It sounds intriguing, but what sort of existence might be possible? It is doubtful that we're even capable of conceiving such a thing. If we could conceive it, then some of us would already be there!

Perhaps it is best to conclude with a reminder that improving the human condition is about more than genetic or technological improvement. Martin Luther King Jr. says it best:

> In spite of these spectacular strides in science and technology, and still unlimited ones to come, something basic is missing. There is a sort of poverty of the spirit which stands in glaring contrast to our scientific and technological abundance. The richer we have become materially, the poorer we have become morally and spiritually. We have learned to fly the air like birds and swim the sea like fish, but we have not learned the simple art of living together as brothers.

As we've all experienced during sickness, biology can make the difference between happiness and unhappiness. Nevertheless, greater happiness and wisdom cannot be reached *only* through better biology. As we explore science, we must attend with equal vigor to the art of being good human beings.

PART V

REINVENTING LIFE

13

THE DAWN OF VIRTUAL BEINGS

Merging Man and Machine

There is no evidence yet that what goes on in living creatures is necessarily different, so far as the physical laws are concerned, from what goes on in non-living things, although the living things may be much more complicated.

—Richard Feynman (*The Character of Physical Law*, 1965)

We have already talked about many radical biological changes in the modern world. In all cases, though, we have thought of life within the traditional confines of birth, death, and DNA. We have made the huge assumption that we know what life is. It seems like common sense, after all. But in truth, we think we know what life is because we've never been challenged to discriminate. The human–machine interface will soon challenge our fuzzy-minded definitions of "life," as well as "death," "intelligence," "consciousness," and what it means to be "human."

Have you ever considered how much time you spend interacting with machines on a daily basis? A typical person might spend several hours on a computer, a few more watching television, thirty minutes driving a car, another half an hour on a cell phone, and so on. If you add it all up, most of us are directly and consciously interacting with machines more than half of the time we are awake. But why?

The simple answer is that machines do things that our biology doesn't allow us to do. The biological world is full of things that humans can't do—we can't spin a web like a spider, survive the frigid cold like penguins, use sonar like bats, or generate our own food like a plant. But with machines, almost anything is possible. We can't fly like birds, but we can use airplanes and helicopters. We can't run as fast as a cheetah, but we can drive cars and ride bicycles to move faster and farther. We can't breathe underwater like fish, but we can use scuba gear and submarines. Machines allow us to talk to each other over great distances (i.e., cell phones and other devices) and control objects without touching them (i.e., remote controls). Machines allow us to see objects that are tiny, distant, or otherwise invisible (i.e., microscopes and telescopes), and even see through objects (i.e., X-rays and thermal imaging). In summary, machines give us superhuman abilities. Think about it: Many of yesterday's comic book and movie superpowers are now possible using today's technologies.

Judging from the amount of time we spend with our machines, we clearly like having superhuman abilities. Is it possible that our technologies will start to become us, and we them? Our egos tempt us to view ourselves as separate and fundamentally different from machines. Nevertheless, as Richard Feynman suggests in this chapter's epigraph, there is no evidence that humans are magical beings. We follow the same physical laws of nature as everything else.

Ironically, not being special is precisely what allows us to have superhuman capabilities. If human bodies and machines obey the same physical laws, then they can be merged. There is no need to rely on theories to make the point—the human–machine interface is already a reality.

Replacing Human Parts with Machine Parts

Almost any human body part that you can think of has been replaced with machine parts for medical purposes. Prosthetic arms and legs are common among war veterans, and artificial hips and teeth are common among the elderly. Cochlear implants allow hearing, and artificial hearts and pacemakers allow blood to flow. Even eyes can be replaced. Researchers at the U.S. Department of Energy and elsewhere have developed bionic eyes—retinal implants that allow sight in people who have lost it due to disease. Finally, using artificial parts for cosmetic reasons is also common, with noses, breasts, and many other parts being replaceable.

Usually the goal of using artificial parts is to restore "normal" human function. Someone without legs is most concerned with walking again. However, artificial parts can carry enhanced characteristics. As we already pointed out, the whole point of most machines is to extend our biological capabilities. A runner named Oscar Pistorius (nicknamed "Blade

Runner") made national news for winning races on two artificial legs. His biological legs were amputated when he was only a year old. His running "legs" have curved pieces of metal on the bottom that act like springs, bouncing him forward with every step. Likewise, researchers are working on retinal implants that allow enhanced vision. Seeing UV light, heat, and other parts of the light spectrum are within reach. Even the X-ray vision of comic books is plausible in the future, as is the transplantation of any other form of imaging technology that exists and can be miniaturized.

Imagine for a moment if we had different goals when we set about replacing body parts. What if our goal was optimization instead of normalization? When someone loses an arm, what if we replaced it with the very best appendage possible, something better than a normal arm? Perhaps the new appendage looks like an arm, but is ten times stronger, can bend in any direction, and has an internal computer with all the functionality of a smart phone. Or perhaps some people would prefer a different type of appendage altogether. Potential enhancements at the human–machine interface certainly get one's imagination stirring. Such enhancements have been science fiction for a century, but only recently have they become technologically possible.

Brain–Machine Interfaces

Even more radical developments lie at the interface between brains and machines. BMIs, or brain–machine interfaces, are allowing electrical activity in the brain to be detected by tiny sensors and then used to control robotic parts. For example, researchers at Duke University, the University of Pittsburgh, and elsewhere have connected the brains of monkeys to computers so that the monkeys can operate a computer screen just by thinking. More recent experiments show that monkeys can perform complex tasks with their mind—like moving a robotic arm to eat fruit. As a proof of concept, the researchers sent the digitized brain signals over the Internet so that the monkeys could control a robotic arm hundreds of miles away in real time. Similar experiments have been done with humans so that they can move wheelchairs, type, operate prosthetic devices, play video games, and do other mechanical tasks just by thinking.

It is difficult to appreciate the profound implications of these experiments. Humans can control machines just by thinking—imagine the possibilities.

Imagine coming home one day in the future after neural connections to mechanical devices have become common. Your car ride home is effortless because you drive just by thinking. When you arrive home, you step out of your car and think the car locked before thinking your house door unlocked and open (keys and door openers are no longer needed). Then you think the lights on and mentally tell your house robot to do the laundry. You sit down to relax in your chair, close your eyes, and connect to the neural Internet. You surf the Internet for a while just as you would today, except that you don't need a computer screen because you can see the screen in your mind. You then contact a friend who is also using the Internet and share thoughts and funny stories. Before going to bed, you live television for a while (you don't "watch" television any more, because now you are mentally embedded in the experience as if it were reality). Finally, after you lie down in bed, you look up at the stars and relax. You can do this because there is a huge telescope in Peru that can be remotely operated across the neural Internet. You look at spiral galaxies and cosmic dust until you drift off to sleep.

As noted previously, it has already been proven that brains and computers can directly interact. It follows that two human brains can interact through a computer interface. The full extent of the possibilities cannot be known at this point. At a minimum, we will be able to do the equivalent of "instant messaging" in our heads and work jointly on projects from a distance as if everyone on a team were physically together. But neural Internet is likely to extend well beyond that. Machines could (theoretically) also interpret unspoken thoughts, emotions, memories, and virtually anything else going on in the brain. Instead of looking at pictures from a friend's vacation, you could look directly at their memories. Or better yet, friends could contact you while on vacation and let you see and experience their vacation with them.

In his famous book *What Is Life?*, Erwin Schrödinger argues, "Consciousness is never experienced in the plural, only in the singular." Perhaps for now. But at the human–machine interface, unique forms of "multiconsciousness" may well be possible. The telepathic abilities of science fiction might not be fictional in the future. In fact, they might be normal.

Consider brain-computer interface experiments along with the artificial eyes being developed at the U.S. Department of Energy. Although vision currently requires a hard-wired connection to a person's optic nerve, the eye itself doesn't have to be located on a person's head. Theoretically, you could see in one place and be in another. Imagine the dual implications of being able to move mechanical parts with your mind and see places without being physically present. Would you like to be an airplane? Or a satellite? Or a robot at a distant location? Each of these options has astounding applications. Maybe you could fight a war without the risk of being at the battle scene. Or maybe you would like to visit a museum on another continent by renting control time for a robot there. Or maybe you want to explore the oceans or go to Mars. Traditional boundaries of the human condition may not be boundaries for much longer.

For many decades, there have been at least two competing visions of what future "enhanced" beings might be like. One involves starting with robots and giving them human-like characteristics to the point that they seem (or are) living. This vision is pursued in the fields of artificial intelligence and advanced robotics. The other vision is moving in the opposite direction — starting with humans and endowing us with capabilities of machines or other organisms. From mid- to late twentieth century, it looked like the first vision was likely to become reality before the second. However, it turned out to be quite difficult to get robots to do even the simplest of human tasks. Higher consciousness for machines still seems a remote dream, at this point. On the other hand, the second vision is an emerging reality on both cultural and scientific levels. We have all embraced machines to the point that they dominate our personal lives, and now science is at a point where we can begin to merge with machines if we choose.

It may be that the machine-to-human versus human-to-machine technological race is a key evolutionary arms race that we won't fully appreciate for several hundred years. If you had to pick a future, would you prefer enhanced cyber-humans or conscious robots? The victors will write the history, as always.

Radical Futures of Human–Machine Interfaces

The advancement of the human-machine interface raises two very interesting and important questions. First, how much of the human body *could* be replaced? With current technology, we clearly could not just take a human head and stick it on a robot body. There are many technological hurdles to overcome, most notably how to keep the brain healthy in a largely artificial environment. Nevertheless, we can say with confidence that most of the body could be replaced without destroying brain function. We know that much from experience.

The second question is how much of the human body *should* be replaced? We marvel at the humanity of restoring livelihood to a wounded war veteran or car accident victim. But is there a point at which replacing parts creates a nonhuman organism? Imagine replacing a human's parts, piece by piece, beginning with the arms, legs, major organs, and so on. How far can you go before you are no longer looking at a human? Most people would answer "the brain," but the answer isn't that simple. What part of the brain makes us human? What if we replaced the entire body and the parts of the brain that control involuntary bodily function? Is the person still a human? Perhaps more important, does having a precise definition of humanity help us? Maybe such definitions are just symptoms of an irrational fear of change. Maybe they are self-limiting, irrelevant limitations that would stifle our next stage of evolution. Or maybe not.

The final destination of technology at the human–machine interface may well be elimination of the body itself. On a visceral level, this might seem repulsive and amoral. On the other hand, both Western and Eastern thought have long histories of minimizing the importance of the flesh. For example, many Christians and Muslims have shunned the human body as sinful to display and insignificant in relation to higher existence (i.e., a soul). On these grounds, one could argue that elimination of the human body altogether would amount to mental and spiritual improvement. As Beatrice says in Dante's *The Purgatorio*, "Control your tears and listen with your soul to learn how my departure from the flesh ought to have spurred you to the higher goal."

In any event, allowing yourself to imagine existence as a floating consciousness in a sea of technological possibilities is captivating. Brains could be paired permanently with mechanical devices like robotic bodies or spacecrafts, or they could just "live" digitally through a computer connection. From exploring the universe to becoming virtual beings, these possibilities represent new frontiers of existence.

It is a dull intellect, indeed, that will not dare to imagine what the future might hold. Possibilities by the end of the twenty-first century include the widespread adoption of computer interactions that feel like reality, bionic vision (i.e., contact lenses with electronics), controlling everyday machines with our minds, neural Internet, conversations without speaking, the end of physical disabilities (for those who can afford it), and humans who can live for hundreds of years. Human nature may change, as well, although such incremental changes are likely to go unnoticed by future generations.

As the organism–machine interface continues to evolve, it is probably inevitable that our definition of life will evolve with it. In fact, our whole concept of life might disappear, just as so many fuzzy ideas of ancient peoples faded away over time. Once the distinction between living and nonliving is sufficiently blurred, existential definitions might simply become irrelevant. If your mother were a cyber-human, how would you define life?

From a scientific perspective, we are clearly a little ahead of ourselves in dreaming of virtual beings and other such things in the early twenty-first century. Therefore, let us finish by returning to what we know. For the moment, we know that the human–machine and human–computer interfaces are real. We have early success stories, and all trends suggest that they are going to improve rapidly. Finally, we know that human bodies obey physical laws like everything else, so there is no physical limitation to how much human–machine merging can occur.

We are left with the likely, yet uncertain possibility that human evolution is about to make a tremendous leap.

14

COSMIC EVOLUTION

Spreading Life across the Galaxy

Since, in the long run, every planetary
civilization will be endangered by impacts from
space, every surviving civilization is obliged to
become spacefaring—not because of exploratory
or romantic zeal, but for the most practical
reason imaginable: staying alive. . . . If our long-
term survival is at stake, we have a basic
responsibility to our species to venture to other
worlds.

—Carl Sagan (*Pale Blue Dot*, 1994)

We have a moral obligation to ensure that humanity
survives in the long run. This being the case, we have a
fundamental problem. We are Earthlings who evolved
to be well adapted to our home planet's conditions.
However, it is irrefutable that the Earth's future is
finite.

Every century brings an uncomfortably high
chance of some cataclysmic cosmic event, like a giant

meteor strike. Even more likely is that we may ruin the planet ourselves through extreme climate change or nuclear war.

If we end up being both lucky enough to avoid cosmic catastrophes and wise enough to avoid wrecking the planet, then the Earth's future is still limited. Our sun will eventually become a red giant thousands of times more luminous than it is today. At that point, several scenarios are possible: the Earth may be scorched into lifelessness, sucked up into the sun and obliterated, or pushed out of orbit to become a lifeless ball of ice. Whatever the case, the natural course of our solar system's evolution will result in the Earth being unlivable for humans (in our current form) in one to seven billion years.

If we could, for just a moment, think about the big picture of our cosmic evolution, then we immediately see that we have virtually no chance for survival in the long run if we remain solely on Earth. It is literally species suicide not to colonize other worlds. In fact, it is even worse—it is Earthicide, as we would doom nearly all of Earth's life to the same fate.

If we accept our role as stewards of life, then part of our responsibility must include spreading life across the galaxy. Although a daunting challenge, all the methods for reinventing life that we have examined provide us with an excellent toolkit. Yes, rapid climate change on Earth is mostly destructive, but it is teaching us how the climates of other worlds could be changed. Yes, we have driven natural selection and genetically engineered organisms with good and bad consequences on Earth, but these things are teaching us how to adapt life for other worlds. Yes, regenerative medicine and the human-machine interface have unpredictable consequences in terms of culture and overpopulation, but they could empower long-term space travel and allow fulfilling existences in space that we never thought possible. Reinventing life on Earth may be just the beginning of our cosmic evolution.

Our Window of Opportunity

The key word of that last sentence was *may*. Colonizing the extreme environments of other worlds will require a synthesis of our knowledge of how to change biology and manage evolution, as well as advancements in engineering. These scientific developments are entirely realistic and achievable. The bigger question is whether we have the will to do so. We probably have a finite window of time in which we have both the technological ability to spread life to other worlds and the societal stability to pull it off. If we as a civilization have the will to spread life during this window of time, then we may survive indefinitely. If we do not have the will and remain self-absorbed, then eventually we will perish.

There is no question that we are currently in such a narrow window of opportunity. We have no way of knowing how long it will last. Two hundred years? Two million years? Whatever the case, now is the time to do the necessary experiments, commit the resources, and take the risks.

This was part of the spirit that propelled space exploration in the 1960s. Yuri Gagarin became the first human to enter space on April 12, 1961, and then Neil Armstrong became the first person to walk on the moon on July 20, 1969. The future of human space exploration looked bright at that point. A total of twelve men walked on the moon in a three-year period. On the Apollo 17 mission, Gene Cernan and Harrison Schmitt spent more than three days on the moon, spending twenty-two hours exploring the surface and covering twenty-two miles using a lunar rover. NASA Administrator Thomas Paine predicted that if we chose, we could assemble a permanent manned space station for perhaps a hundred scientists by the mid-1970s, a base on the moon that could be occupied for months by the late-1970s, and a manned journey to Mars by the 1980s. He was not exaggerating; we could have done all of this. But we didn't. Humans haven't strayed far from Earth since Apollo 17, and the *last* people walked on the moon on December 11, 1972.

Meanwhile, the cosmic clock is ticking on our window of opportunity for spreading life to other worlds. Tick tock, tick tock.

On the positive side, space exploration and research over the past forty years have generated a wealth of new knowledge and practical applications for societies on Earth. We've sent satellites and robots to the far reaches of the solar system, peered back to the origins of the universe, built an international space station, driven a remote-controlled rover to explore Mars, and done thousands of experiments in space. Likewise, the list of technologies that were made possible by space research is stunning—satellites, cell phones, prediction of weather and natural disasters, medical imaging, protective clothing, smoke detectors, cordless tools, insulating materials, advanced plastics, joystick controls, and thousands more. The more we explore space in the future, the more we will discover that allows us to solve problems on Earth. As Philip

Scranton concludes in analyzing spaceflight's impact on society, "Beneath the satellites, probes, and human spaceflights, for a generation or more extensive innovations in process, materials, and instrumentation have flowed outward from NASA projects and resonated through the industrial economy."

Refocusing on the goal of humans living in space will have an even greater positive impact on Earth. Colonizing space will challenge us to address many of the problems we have here on Earth in new, creative ways. For example, a space colony would have to recycle virtually everything—water, air, energy, and basic materials. How can we most efficiently convert urine and waste water back into drinkable water? How can we balance energy generation with energy use in a sustainable way? How can we generate food in the most efficient way possible? Thinking about these issues in the context of space has direct applications on Earth in terms of new crops, recycling methods, greater efficiency, and renewable sources of energy.

The practical benefits of space exploration and colonization far exceed the resources and risks to pursue them. The larger benefits for the human condition are truly incalculable.

The Effects of Space on Earth's Life

Imagine that you are on a spaceship containing a variety of different species. You have other animals, plants, bacteria, and fungi for experiments and food. The engines rumble as you lift off from Earth and enter the blackness of space. From the first instant in space, every species in your spaceship changes. . .

Let's start with humans. The most obvious immediate impact is that we no longer feel the effects of Earth's gravity. The floating is fun, but makes half of us sick for a few days until we get used to it. The other immediate impact is our fluid balance. Without gravity, blood distributes itself evenly throughout the body, resulting in more fluid than usual in the upper part of the body. Our heads are a little puffy and congested, sort of like the feeling of having a cold.

As the days and weeks pass, we experience many physiological decreases: we lose weight, our muscles atrophy, we lose calcium and potassium, our bone strength is reduced, and our red blood cell count goes down. Most of these effects sound more dire than they probably are. Our bodies adapt to the local environment, and when we weigh nothing, big muscles and bones just aren't that important. Nevertheless, there are countermeasures we could use to reduce all of these effects, such as increased exercise, better nutrition, and centrifugation (spinning to simulate gravity).

Among the biggest long-term concerns for humans and other organisms in space is radiation. As we discussed earlier, increased radiation causes DNA mutations, tissue damage, and sometimes death. For example, rats exposed to several years' worth of space radiation have been permanently impaired. Likewise, the amount of radiation that a human would encounter in deep space over several years is likely to be debilitating and/or deadly. In 1969, Buzz Aldrin was the first to experience was is now called a "retinal flash," a perceived flash of light after a radioactive particle rips through one's head. Ultimately, such events could have serious health consequences.

The current record-holder for space longevity is Valeri Polyakov, who spent more than fourteen consecutive months in Russia's Mir space station and a total of twenty months in space. He returned to Earth in very good health. However, current space stations are still within Earth's magnetosphere, shielding the astronauts from the worst effects of radiation. There is great uncertainty about how quickly humans would succumb to the radiation of deep space. Nevertheless, some researchers have hypothesized that between 13 and 40 percent of an astronaut's neurons would be struck by cosmic rays over the time it would take to travel to Mars and back. In other words, unless we figure out how to block the radiation or fix the damage, a trip to Mars might be a death sentence. We return to the radiation problem in a few pages.

There are also substantial psychological stresses in space. Spaceships and space stations so far have been relatively small and sterile. Imagine being locked in a small space with several other people for years, with no escape during times of conflict. The nature of this stress isn't unique to space, but is typical of all small, closed environments. Similar stresses have been documented on small boats, submarines, jails, caves, and distant research stations in Earth's polar regions.

We can imagine the larger and more permanent space colonies of the future, but it is difficult to predict what the long-term psychological effects of living there would be. Nevertheless, given all of the environments that humans have lived in over the millennia, it certainly seems possible that we would be fine. This is especially true for the future generations born into space colonies, who would have no other frame of reference.

Other organisms change in response to the space environment, as well. For example, plants grow differently in space than they do on Earth. In the absence of gravity and without an "up" or "down," plants lose sense of which direction they should be growing. They grow in a mottled mess if you don't provide simulated gravity by spinning them or placing a magnet nearby. They also tend to photosynthesize less and produce less cellulose and lignin, the materials in cell walls that make plants stiffer. Plant reproduction and germination remain viable in space as long as they are grown with sufficient air flow, water, and nutrients. Overall, the process of growing plants in space is still a lot like growing them on Earth—if you play around and try different growing conditions, plants can thrive.

Obviously, it is important that we figure out how to grow plants and other organisms in space if we hope to create sustainable colonies. Plants can provide nearly everything humans need—food, clean water, oxygen, fiber, and medicines. They are the ultimate life support systems.

To explore and colonize space, we need a clever long-term plan for dealing with all the biological changes that will affect us. Looking back across the history of human exploration and colonization, we see two trends of biological change.

First, we see humans transforming their new environment to suit their needs. Colonists bring seeds, animals, diseases, and horticultural practices that push back native life and allow humans to move in. Second, we see the adaptation of humans themselves, along with the adaptation of the other organisms that humans bring along. For example, we see skin color adapting to the amount of sunlight, immune responses adapting to new diseases, and digestive systems adapting to nutritional sources. Crops and livestock are also adapted, whether consciously or unconsciously, to fit the selective pressures of new environments. In space, it seems inevitable that we will see both of these trends over and over again, changing environments to suit life and changing life to fit new environments. How might this happen?

Altering Other Worlds to Fit Earth's Biology

We have already transformed planet Earth in a variety of ways. Thus, if we choose, we can alter other worlds so that they are more suitable for Earth's life. This is often referred to as "terraforming." Ultimately, this means making other worlds more like the modern Earth.

As would be immediately obvious to any alien visitor who saw Earth for the first time, this is a planet of water. Water is absolutely essential to life here—it makes up a large proportion of our biological systems, creates the atmosphere, and buffers the entire planet. Any world that will be hospitable to Earth's life must have water. Once we have identified a world with water, the next important environmental variable is temperature, because it is the key to providing the water in a liquid form. Once water is a liquid, a water cycle can develop through the atmosphere, the world's surface, and whatever species we bring with us. This is likely to be the tipping point for life in many worlds— the point at which life grabs a foothold and retains it, converting barren landscapes into vigorous living biospheres that evolve on their own.

Let's compare Earth and Mars as an example. On the positive side, Mars has soil that is suitable for growing plants, as well as abundant water. However, Mars is colder (-75°C to 10°C), has about a third of the gravity, and has less than 0.1 percent the atmospheric pressure. The composition of the air is also different, with more than twenty-five times the amount of carbon dioxide but with hardly any oxygen or water vapor. Thus, if we wanted to terraform Mars, we would need to increase the temperature, increase atmospheric density and pressure, and alter atmospheric composition.

We might begin the terraforming process with simple life forms, in much the same way life began on Earth. We could start by introducing genetically engineered microorganisms, which would begin building an atmosphere. After the environment is a little more hospitable, targeted plant and fungi species could be introduced. As the environment becomes even more hospitable, insects and other animal pollinators could be introduced. Later, larger animals might be able to survive. Other techniques for terraforming that have been proposed include injecting greenhouse gases, altering Mars's rotational axis, or using space mirrors. It is also possible to start with small space stations or biospheres and progressively build toward transforming whole planets, a process referred to as paraterraforming.

The ideal outcome would be that humans, perhaps with some appropriate adaptations, could walk around comfortably in the environment. It is doubtful that we could ever create a perfect replica of Earth's environments, and it would be naive to think that terraforming would be simple or predictable.

Nevertheless, it could at least bring conditions within a livable range and make life easier for a human colony. Better yet, we might get some ecosystem services from another world's environment (i.e., water cycling, radiation protection, and soil building) just as we do on Earth.

Altering Earth's Biology to Fit Other Worlds

William Burroughs, the twentieth century author, wrote a story about a man named Kim whose sole purpose in life was space travel. Burroughs wrote:

> Kim read all the science fiction books and stories he could find and was stunned to find the assumption, the basic assumption, that there is no real change involved in Space travel: same dreary people playing out the same tired old roles. Take that dead act into Space. Now here they are light years from Planet Earth watching cricket and baseball on a vision screen. . . can you imagine taking their stupid pastimes light years into Space. It's like the fish said, "Well, I'm gonna just shove this aquarium up onto the land and there I've got everything I need. . . (laughter). . . you need entirely too much.

We do need "entirely too much" for most space environments. It may well be true that the only way to successfully colonize space is to adapt ourselves and other organisms to thrive in particular space environments.

Adapting Earth's life to fit other worlds is

completely realistic. We have to keep in mind that our planet has changed radically over the 3-billion-year history of life here. Thus, Earth's life has already gone through many of the so-called extremes that we will encounter in space.

In fact, many of the adaptations needed to live in extreme space environments already exist somewhere on Earth. Life thrives on Earth, everywhere from boiling hot springs to the frigid depths of the Artic Ocean. Biologists have spent decades studying the "extremophiles" that live in these environments to understand how they are able to remain chemically stable. We understand them well enough that they are already used in many industrial processes, which often occur at extreme temperatures, acidity levels, and pressures. Space colonists may find extremophiles and their genes invaluable for establishing a biological foothold in the difficult environments of space.

Another likely scenario is that we will use human–machine interface technology (described in the last chapter). In space, the ability to control machines with our minds would be incredibly helpful. Space is filled with perils. Instead of putting on a spacesuit and entering a dangerous space environment, why not send a robot surrogate through which you can see and touch? Or perhaps you are outside making a repair and you need to access the neural Internet to look at an instruction manual (neural Internet won't float away like a book or computer screen). Or maybe you need biosensors connected to all the organisms of a space station so that they can coordinate their biology, balancing oxygen and carbon dioxide levels, food, energy, and waste. The possibilities are endless.

As we've already seen, the capacity to cope with

radiation is also important in space. There is simply no way to exist in space for very long without blocking the radiation, rapidly fixing the biological damage, or both. The former would require a huge addition of weight to spaceships that is not currently feasible, or perhaps the invention of a portable magnetic force field.

On the other hand, rapidly fixing the mutations and related biological damage may be possible using regenerative medicine and gene therapy. This creates a tremendous opportunity for synergy between space research and Earth-focused research. As discussed in previous chapters, DNA mutations are the root cause of cancers and thousands of genetic diseases. Likewise, DNA mutations in microbes foil our attempts to cure AIDS, malaria, and other infectious diseases. Thus, research and experiences related to long-term space exploration contribute to curing some of Earth's worst diseases. Conversely, advances in gene therapy and regenerative medicine on Earth will provide essential tools for future space colonists. Eventually, space-dwelling organisms may be engineered with heightened radiation and mutation tolerance.

Although an increased number of DNA mutations is nearly always bad for individual humans, it can lead to rapid evolution on the level of whole populations. If you imagine human colonies on several other worlds, intriguing possibilities are created over the long run. All the ingredients for rapid and dramatic evolution will be in place — small populations, reproductive isolation, and a rapid mutation rate. Thus, humans in space colonies are likely to evolve more rapidly than humans on Earth.

Evolutionary pressure could possibly solve many of the biological problems of living in space, though at the price of many deaths. Natural selection is not kind, after all, so it would be better to genetically engineer the solutions when possible.

Similar evolutionary pressures will affect all of the organisms we bring with us into space. Because the environment of each world is different, the biological adaptations needed to live on each world will also be different. As on the islands of Earth, biology on other worlds is likely to drift apart over time, eventually creating unique and self-sustained ecosystems with their own symbiosis and ecological balance. For example, we'll need crops optimized to thrive in the light, soils, and waters of different locations. Thus, future life on Earth, the Earth's moon, Europa, Titan, and other worlds could all be radically different. We might also find that engineering humans to breathe different air, tolerate altered gravity, or survive radical temperatures is more practical than altering the air, gravity, or temperature of an entire planet.

The colonization of space may eventually be a turning point for biodiversity. We have driven Earth's life into the sixth great extinction, and it appears that this decline will continue. But over time, the nature of space environments might lead to a new and unique expansion of cultures and species. Perhaps space colonization will be the great turning point where humans become ambassadors of Earth's species instead of its destroyers. It is our choice.

Our Place in the Universe

A quick journey around our solar system would show us hundreds of celestial bodies, from Mercury to the moons of Jupiter, which are essentially lifeless balls of rock or poisonous gas. They are beautiful objects of nature worthy of study. Nevertheless, without life, they serve no real purpose. Just as life migrated across the face of the Earth, transforming itself and the planet in the process, so should we migrate to other worlds.

It may be part of the human destiny to spread life across the galaxy. It might even be our most important role. We are ambassadors for life, or galactic farmers, if you like. It is difficult to think of a more admirable goal for a species than to spread life and intelligence to lifeless worlds.

It is telling that the biggest word we have for broad thinking is *worldview*. That's just not big enough. A broad worldview can only take us so far. What the human species needs next is a *cosmic view* – a view of ourselves as agents of life across the far reaches of space and time. Our mental expansion must precede our physical expansion.

We are just getting started in the universe. Earth is our nursery, and cosmic evolution is our future. We've crawled, taken our first steps, and even said a few words. . . now it is time to walk and talk and move beyond the nursery, for the universe awaits.

15

THE CLEARING FOG

Accepting Our Role as Stewards of Evolution

Lead us, Evolution, lead us
Up the future's endless stair:
Chop us, change us, prod us, weed us.
For stagnation is despair:
Groping, guessing, yet progressing,
Lead us nobody knows where.

—C.S. Lewis ("Evolutionary Hymn" in *Poems*, 1964)

*T*he most exciting and relevant aspect of evolution is that it is still happening—the evolutionary tree is still growing. If we could come back in a million years, much would be unrecognizable. Our food, diseases, pets, and wildlife are all evolving around us and in response to us. Based on more than 3 billion years of historical precedent, we should expect that new species will continue to appear, some will go extinct, and every species will change over time. That includes humans.

As we have found throughout this book, humans are directing evolution in a multitude of ways. We are, in fact, the dominant evolutionary pressure at the present time. We are driving many species to extinction, making a few species very prolific, changing species, and even creating new ones that have never existed before. In most cases, we impact life without concern for the evolutionary implications. Our impacts are often unconscious and haphazard. To go further, our impacts are often ignorant, shortsighted, and destructive. We can do better.

It is time that we accept our role as evolutionary stewards—the future of life is our responsibility. For too long people have argued about whether it is right to change nature. This is a useless argument. We already *are* changing nature in profound ways. Anything we do at this point is going to continue to change it, whether that involves destruction, creation, or restoration. It really doesn't matter whether you like it, whether your faith agrees with it, or whether your science training includes it. Future life will not look like life does now, and human decisions will drive it in whatever directions it goes. We may as well do a good job.

What does it mean to be a good steward of evolution? Specific context is essential for making good evolutionary decisions. Nevertheless, we may generalize five core principles: diversity, sustainability, conscious decision making, improvement of the human condition, and expansion. *Diversity* is necessary for the health of any biological system, as it provides the raw material for adapting and evolving to changing environments. *Sustainability* ensures that we leave our world in a condition that allows future generations to meet their needs. *Conscious decision making*, though imperfect, is far preferable to the random and unconscious ways that we are currently driving evolution. *Improvement of the human condition* means that we continue to strive for progress, seeking higher forms of existence, consciousness, and wisdom. Finally, *expansion* to other worlds ensures the continuance of life beyond the finite future of Earth. These principles would lead us forward to a safe, compassionate, and adventurous future.

Adjusting to Radical Change

In the *Divine Comedy*, Dante describes his final approach to Hell as a gradual revealing:

> Just as a man in a fog that starts to clear
> begins little by little to piece together
> the shapes the vapor crowded from the air —
> so, when those shapes grew clearer as I drew
> across the darkness to the central brink,
> error fled from me; and my terror grew.

This adequately expresses the fears of many regarding the future of life in the fog of radical biological change. Will our journey plunge us into a hell from which we cannot emerge? Or, like Dante, are we ultimately headed for a progressive ascent into a better existence? Obviously, we would prefer the latter possibility.

Make no mistake—radical biological change is coming regardless of how well we prepare ourselves. Undoubtedly, some people will be unprepared and disoriented, just as some are now, whereas others will acknowledge change and thrive. Perhaps the best way to prepare ourselves as a culture is simply to expect and accept change itself.

To appreciate how future generations will

embrace the radical changes of the future, we have to appreciate the mental plasticity of children. Future generations will not be mentally trapped within the twenty-first century like we are. It takes children a while to distinguish real from imaginary and to acclimate to the environment of their historical period. As a result, it takes them a while to figure out what is currently real and what is not.

Consider my son Luke as an example. When Luke was two years old, we decided to take him to the zoo for the first time. After I explained to him that we would see all sorts of animals, he began asking me, "Will we see elephants?" "Will we see monkeys?" and so on. We had been reading all sorts of books, some with modern animals, some with dinosaurs and other species that are extinct, and some with imaginary animals. Therefore, when he asked, "Will we see a brontosaurus?" he was sincerely curious. No answer would have surprised him. The reality of his lifetime and what he would consider "normal" and "natural" were being partly established at that moment. "No, Luke, we can't see a brontosaurus. They are extinct." Perhaps cloning technologies will allow future parents to answer differently, and then "normal" and "natural" will change.

After watching Peter Pan, Luke asked me if he could fly. I answered, "You can fly in an airplane." Not satisfied, he continued, "No, can I fly like Peter Pan?" I thought and answered, "Nobody can fly like Peter Pan right now, but if you're smart enough you might be able to figure it out." Indeed, we can imagine that astronauts of the future, who are already floating in zero gravity, could engineer a small biological or mechanical adaptation to propel themselves around.

Perhaps my favorite question from Luke was when we were reading a book and he saw a picture that he really liked. "I want to go in there," he said, sticking his fingers into the crease of the book and moving his head toward it. "Why can't I go in?" I laughed and said, "You just can't. It is just a picture." He was clearly disappointed. But perhaps future parents will allow their children to wander around in virtual realities that enhance learning. I fully expect that when Luke's children or grandchildren ask him the same question, he'll be able to say, "Sure! Let's go in the book". . . and off they will go. Whatever the future will be, rest assured that radical and profound changes to us will be completely normal to future generations. The mental plasticity of children should be a reminder to us all that virtually anything is possible.

The other side of reality, however, is that most adults loathe change. Yes, people love the *idea* of change and that it often represents progress. But when it comes down to specifics, most people are very cautious and sometimes belligerently resistant to major changes in their lives. The environmental and technological realities of our lifetime trap our minds and become our status quo. The world we are born into becomes our *de facto* definition of "normal" and "natural," arbitrarily shaping our preferences and our ethics. Thus, to consider the future with an open mind, we must acknowledge how biased we are because of when we were born.

People from 400 years ago would view us as having superhuman abilities—we live much longer, we are taller, we are wealthier, we can communicate across the globe in an instant, we can fly into space and traverse the oceans, and we can cure disease. Likewise, future humans may seem superhuman or transhuman to us. From their perspective, we may seem like ancient tribes seem to us today. They may respect and remember us, but they will never turn back. Nor will they be able to imagine how we tolerated many of the inconveniences that we do, like disease, ignorance, and death.

Choosing a Future for Life

Whenever someone claims to foresee the future, it should automatically set off our BS-meter. Nobody knows the exact future. Not horoscopes, not scientists, not the pope, not futurists, not stock market analysts, not biblical literalists, not politicians, not you, and certainly not me. The future results from choices that people make, and until they make them, we can't know exactly how things will turn out. This is a tremendous opportunity. It means that the future is, in fact, in our hands.

Over the course of this book, we've explored many of the big questions that we must answer to shape the future of life. Let us now summarize some of our most important choices. Considering them all at once, though dizzying, allows us to consider how the various pieces of the future may fit together. The synthesis of our answers to these questions will shape the future of life for all time.

1. How much of the world's natural ecosystems will be preserved?
All of the conservation efforts in the world will make no difference unless we set aside large amounts of land

and ocean for preservation. The degree to which we are able to preserve natural ecosystems will determine not only biodiversity, but also the long-term success of human societies. We rely on natural ecosystems for water, air, food, fiber, and a multitude of other things. As the human population continues to swell, will we be able to increase efficiency and maintain a steady-state level of natural ecosystems? Or will we continue to sprawl and degrade the planet? Do we have the courage to restore ecosystems on a regional scale and actually improve the situation?

2. To what extent can we slow climate change?
Climate change is the great wildcard for the next thousand years. Human activities are changing the composition of the atmosphere and thus the climate, with an enormous spectrum of consequences. As climate shifts, societies will be forced to migrate (or deal with migrants), change agricultural practices, and cope with unusual and unpredictable environmental conditions. Likewise, many species will no longer be properly adapted to their environments, leading to species declines and extinctions. Will we control emissions of greenhouse gases to slow climate change? Will we regrow forests and build soils to sequester more carbon from the atmosphere? Will we create natural corridors to allow ecosystems the chance to shift from on region to another? Or will we maintain the delusion that climate change isn't happening, we have nothing to do with it, and we can't fix it anyway?

3. What will be the characteristics of future foods?
We are our food. We noted earlier in this book that most modern foods in the United States and elsewhere

have been selectively bred and then genetically engineered. If we traveled back in time a few thousand years, almost all of our food would look very different—some of it would be unrecognizable. It follows, then, that food will continue to evolve if we choose to direct it. Should future food be bred and engineered to be healthier and to taste better? How must farming systems adjust to cope with rapid climate change and harsh conditions such as drought, salty soil, flooding, extreme heat and cold, and pathogen attack? Likewise, how must farming and eating change to reduce environmental impact? How do we ensure genetic diversity within our increasingly industrialized food system? How can we balance food consumption to reduce both the hunger of developing countries and the overindulgence and rising obesity of developed countries?

4. How far will we extend human life spans?

Life spans have doubled over the past few hundred years in the developed world, a remarkable achievement of public health and technology. With cures for existing diseases, we could increase life spans by perhaps another twenty to thirty years. With new technologies such as regenerative medicine, gene therapy, and the human–machine interface, extending life for hundreds of years may be possible. Increasing life spans would transform much about how families and societies function. What are the costs and benefits of increasing life spans? How do we keep from overpopulating the planet as life spans increase? How will families and communities change with more and more older people (and possibly fewer younger people)? Would being older make us wiser, or would it stifle progress?

5. Will human reproduction be driven by randomness or intentionality?

Sexual reproduction is inherently unpredictable. Genetic material from both parents comes together randomly to form new combinations, creating a unique life in the process. The problem is that genetic mutations and chromosomal abnormalities often occur, leading to sickness, suffering, and death. Likewise, mutations occur throughout our lives, several hundred of which are typically passed on to our children. Genetic tests and therapies have the potential to both eliminate diseases and alter our evolutionary path. Should we use genetic screens and gene therapy to eliminate deadly diseases? Could such methods reduce miscarriage and abortion rates? Should family planning include choosing gender or deciding to have twins? Should we eliminate some genetic diseases from our gene pool entirely?

6. Will we limit the genetic enhancement of physical traits? If so, how?

Our genes determine a great deal about who we are as individuals and as a species. As we learn more about our own genetics, we'll have more opportunities to enhance strength, endurance, metabolism, vision, intelligence, beauty, and other characteristics. The stakes are huge. What choices would people make if given complete freedom? Will those who choose not to participate be outcompeted? How might the military enhance soldiers, and might a new arms race develop from this? Will new sports or categories within sports emerge for those who are genetically enhanced? Will increasing intelligence allow us to solve global problems more easily and bring widespread prosperity, or instead have unforeseen negative consequences? Could more radical human traits be engineered, perhaps even leading to new human species?

7. How will humans merge with machines?
We have already replaced many human parts with manufactured machine parts. In addition, computers have been successfully connected with brains, allowing human thoughts to use computers and control machine parts. How will we deal with situations where replacement parts give people superhuman abilities? How will our brains adapt to internalizing activities that are now external, such as using the Internet, playing video games, and communicating with others? Is there a point in the merging of humans and machines beyond which we lose our humanity? And should we care? Or should we forge ahead and experiment with transhumanism?

8. What new forms of life will be created?
New fields such as synthetic biology have the potential to create entirely new species, dramatically changing the course of evolution. It will begin with genetically engineered microorganisms and then progress to more complex species. It may also include the creation of entirely new forms of life, which may or may not be DNA-based. How might new species be useful for solving global problems? How might new species interact with existing species? Might they endanger existing life? Are there intelligent life forms that we could create, perhaps at the boundaries between biology and computers?

9. When and how will we spread life to other worlds?
Deciding to colonize other worlds could lead to enormous evolutionary steps. Currently, there is an organism–environment mismatch between Earth and other worlds. But it is certainly not an insurmountable mismatch. In the future, the gap can be closed by engineering Earth's organisms to fit other worlds and vice versa. Once life is established on another world, it would then begin to evolve on its own. We Earthlings may well become galactic farmers who sow the seeds of life all across the Milky Way. Or, if we decide differently and stay on Earth, evolution will take a completely different course and ultimately will end as the Earth becomes uninhabitable. Should we colonize other worlds? If so, how high on our priority list should space exploration be? Is it our role to spread life across the galaxy? Will we achieve a "cosmic view" as a species?

10. What will give humans the wisdom to direct evolution responsibly?
This question might seem like a rhetorical throwaway, but it underlies how we approach every other question related to the future of life. When confronted with unprecedented challenges and choices, what will be the basis for people's decision making? Will we use science, reason, and critical thought? Or will we revert to shallow ideologies and political partisanship? Or will we rely on faith traditions and ancient doctrines? Will we accept that change is a part of the world and envision a better future? Or will we allow ourselves to be trapped in the status quo of the present day? How will we change our educational system for a world of rapid and radical biological change? Should we strive

for progress and, if so, what does "progress" mean?

Evolutionary Stewardship

It might be tempting to think that the future of life will be determined by an unconscious momentum of society or by the typical forces of evolution. This is absolutely not the case. The future of life will be determined by human decisions—our decisions. We have choices. The degree to which we accept our roles as evolutionary stewards and govern responsibly will determine our own fate and our legacy for future generations.

Innovations in science and technology create new opportunities for progress, but they do not guarantee progress. For real progress to occur, innovations must be accompanied by other advances in society. This second round of innovation usually involves people who are scientifically literate but are not scientists themselves. They figure out how to use scientific ideas and new technologies to solve problems in their communities, create economic opportunities, and better educate people. They use science as a tool to improve the world, including humanity and the rest of Earth's species. Maybe you can be one of these people. I sincerely hope so.

I have faith in humanity. I have confidence that when people in a free society are well informed, focus on the right questions, and look to the future, they can make good decisions that will improve the lives of future generations. So I am optimistic that if we embrace the idea of evolutionary stewardship, we will find a way to a better future.

Evolutionary stewardship is a necessary outcome of being the dominant species on our planet. Just as humans once roamed the Earth freely without governments until governments were necessary to prevent chaos and destruction, so we have reached the point where evolutionary governance is necessary to prevent chaos and destruction. We must direct biological change because not directing it would allow change to spiral in wild and perilous directions. Likewise, directing evolution could reduce human suffering, improve quality and quantity of life, and offer transformative possibilities that we can hardly comprehend right now.

Evolutionary randomness and recklessness have become, and hereafter shall forever be, unethical and unacceptable. We must design the future of life, piece by piece, and with great compassion and care, so that life on Earth can be sustained and eventually spread to other worlds.

ACKNOWLEDGEMENTS

I owe my most sincere and loving thanks to my wife, Beth, who was so supportive during the whole process of writing this book. She was the first to read every chapter, usually giving advice that made the text more readable for everyone else. She also participated in many life experiments that resulted from thinking through the book, such as raising backyard dinosaurs (chickens), searching for bizarre human mutations in San Francisco, setting off homemade rockets, and embracing obsessive energy conservation measures.

Likewise, my sons Luke and Vince were incredibly patient on many occasions as Dad needed "just a minute" (an hour) to work on this book. My passion for the future of life is undoubtedly heightened by my love for them.

I am indebted to the many people who provided feedback and criticism on my manuscript, including David Vandermast, Bud Warner, Tony Crider, Steve Braye, Mary Williams, Mat Gendle, Peter Felten, Deborah Wells, Greg Haenel, and Eric Townsend.

Two of my colleagues at Elon University were especially helpful and gave critical feedback on the entire book. First, Jean Schwind provided a brilliant literary review that improved the manuscript in a variety of ways. Second, Dave Gammon offered spirited and challenging criticisms that led to several important improvements.

My sister-in-law, Laura Poole, provided her excellent copy editing eye to put the finishing touches on the manuscript.

The Center for the Advancement of Teaching and Learning at Elon University supported the early development of my Reinventing Life course. Much of the research and development of this book took place as a result of that support.

The research and ideas of many scientists contributed to this book and I am indeed a "midget standing on the shoulders of giants," as the saying goes. Works by Stephen Palumbi (Stanford University) were especially influential in the early stages of this project.

I'd like to thank the hundreds of students in my Reinventing Life and Global Experience classes at Elon University. Their creativity and critical thinking inspired many directions that this book takes, and cautioned against several that it doesn't. Every day I am inspired by the opportunity to interact with such wonderful students.

Finally, I'd like to thank Sophia Clotho for her wisdom and inspiration as we thought through our contributions to the grand conversations of humanity.

The many sources and reviews of this text were enormously helpful and shaped the book's language in a variety of ways. Nevertheless, I accept full responsibility for any weaknesses or errors.

REFERENCES

From Jaguars to Avatars
Palumbi, S. 2001. Humans as the world's greatest evolutionary force. Science 293: 1786-1790.
Sagan, C. 1977. The dragons of Eden. New York: Ballantine Books.

Evolution All Around Us
Darwin, C. 1859. The origin of species. New York: Barnes and Noble Classics, 2004.

Our Greatest Irony
Carpenter, K.E. et al. 2008. One-third of reef-building corals face elevated extinction risks from climate change and local impacts. Science 321: 560-563.
Halpern, B.S. et al. 2008. A global map of human impact on marine ecosystems. Science 319: 948-952.
Jergensen, C. et al. 2007. Managing evolving fish stocks. Science 318: 124.
Lotze, H.K. 2006. Depletion, degradation, and recovery potential of estruaries and coastal seas. Science 312: 1806-1809.
Lubowski, R.N., Vesterby, M., Bucholtz, S., Baez, A., Roberts, M.J. 2002. Major uses of land in the United States, 2002. Economic Research Service/USDA. Document 2002/EIB-14.
Lyell, C. 1830. Principles of geology. London: Penguin Books, 1997.
Millennium Ecosystem Assessment. 2005. Ecosystems and human well-being: Synthesis. Washington,

D.C.: Island Press.

Pandolfi, J.M. et al. 2003. Global trajectories of the long-term decline of coral reef ecosystems. Science 301: 955-958.

U.S. Department of Energy. 2000. United States nuclear tests – July 1945 through September 1992. DOE/NV-209-REV 15. Nevada Operations Office, Las Vegas.

Forgotten Worlds

Huxley, J. 1942. Evolution: The Modern Synthesis. New York: Harper.

International Union for Conservation of Nature. 2010. The IUCN Red List of Threatened Species. Version 2009.2. www.iucnredlist.org.

Lawson, J. 1709. A new voyage to Carolina. Chapel Hill: UNC Press, 1967.

Lewis, C.S. 1964. Poems. San Diego: Harcourt.

Mendelson, J.R. et al. 2006. Confronting amphibian declines and extinctions. Science 313: 48.

Millennium Ecosystem Assessment. 2005. Ecosystems and human well-being: Biodiversity synthesis. Washington, D.C.: World Resources Institute.

Roberts, C. 2007. The unnatural history of the sea. Washington, D.C.: Island Press.

Schipper, J. et al. 2008. The status of the world's land and marine mammals: Diversity, threat, and knowledge. Science 322: 225-230.

Nature or Nothing

Club, R., Rowcliffe, M., Lee, P., Mar, K.U., Moss, C., and Mason, G.J. 2008. Compromised survivorship in zoo elephants. Science 322: 1649.

Enserink, M., and Vogel, G. 2006. The carnivore

comeback. Science 314: 746-749.

Gibbons, A. 2010. Greening Haiti, tree by tree. Science 327: 640-641.

Jackson, T.J., and Hobbs, R.J. 2009. Ecological restoration in the light of ecological history. Science 325: 567-568.

Lamb, D., Erskine, P.D., and Parrotta, J.A. 2005. Restoration of degraded tropical forest landscapes. Science 310: 1628-1632.

Mora, C., et al. 2006. Coral reefs and the global network of marine protected areas. Science 312: 1750-1751.

Morell, V. 2008. Wolves at the door of a more dangerous world. Science 319: 890-892.

Morell, V. 2008. Into the wild: Reintroduced animals face daunting odds. Science 320: 742-743.

Nepstad, D. 2009. The end of deforestation in the Brazilian Amazon. Science 326: 1350-1351.

Schweitzer, A. 1949. The philosophy of civilization. Amherst, NY: Prometheus, 1987.

Willis, K.J. 2006. What is natural? The need for a long-term perspective in biodiversity conservation. Science 314: 1261-1265.

Wilson, E.O. 2002. The future of life. New York: Vintage Books.

Wuethrich, B. 2007. Reconstructing Brazil's Atlantic Rainforest. Science 315: 1070-1072.

Size Matters

Alberts, B. 2008. The promise of cancer research. Science 320: 19.

Coltman, D.W., et al. 2003. Undesirable evolutionary consequences of trophy hunting. Nature 426: 655-658.

Coltman, D.W. 2008. Molecular ecological approaches to studying the evolutionary impact of selective harvesting in wildlife. Molecular Ecology 17: 221-235.

Diamond, J. 1997. Guns, germs, and steel. New York: Norton.

Gill, S.R. et al. 2006. Metagenomic analysis of the human distal gut microbiome. Science 312: 1355-1359.

Krajick, K. 2003. Methuselahs in our midst. Science 302: 768-769.

Tuomanen, E. 2005. Appreciating Our usual guests. Science 308: 635.

When Evolution Turns Deadly

Bloland, P.B. 2001. Drug resistance in malaria. World Health Organization. WHO/CDS/CSR/DRS/2001.4.

Diamond, J. 1999. Guns, germs, and steel. New York: Norton.

Laufer, M.K., and Plowe, C.V. 2004. Withdrawing antimalarial drugs: Impact on parasite resistance and implications for malaria treatment policies. Drug Resist. Updates 7: 279-288

National Institutes of Health (National Heart, Lung and Blood Institute). 2007 (Nov.). Sickle cell anemia. www.nhlbi.nih.gov/health/dci/Diseases/Sca.

Palumbi, S.R. 2001. Humans as the world's greatest evolutionary force. Science 293: 1786-1790.

Plowe, C.V. 2003. Monitoring antimalarial drug resistance: Making the most of the tools at hand. Journal of Experimental Biology. 206: 3745-3752.

Sabeti, P.C. et al. 2006. Positive natural selection in the human lineage. Science 312: 1614-1620.

Stearns, S.C., and Koella, J.C. 2008. Evolution in health and disease. 2nd ed. New York: Oxford University Press.

Taubes, G. 2008. The bacteria fight back. Science 321: 356-361.

World Health Organization (WHO). 2004. The world health report 2004 — Changing history. www.who.int/whr/2004/en/index.html.

Growing a New You

Gould, S.J. 1977. Ever since Darwin. New York: Norton.

Gurdon, J.B., and D.A. Melton. 2008. Nuclear reprogramming in cells. Science 322: 1811-1815.

Holden, C., and G. Vogel. 2008. A seismic shift for stem cell research. Science 319: 560-563.

Sandburg, C. 1953. Honey and salt. New York: Harcourt, Brace & World.

Vogel, G. 2008. Reprogramming cells. Science 322: 1766-1768.

Wake Forest Institute for Regenerative Medicine. 2010. http://www.wfubmc.edu/WFIRM/.

Clones Clones Clones

Cibelli, J. 2007. A decade of cloning mystique. Science 316: 990-992.

Folch, J., et al. 2009. First birth of an animal from an extinct subspecies (Capra pyrenaica pyrenaica) by cloning. Theriogenology 71: 1026-1034.

Fulka, J. Jr., Loi, P., Ptak, G., Fulka, H., and St. John, J. 2009. Hope for the mammoth? Cloning and Stem Cells 11: 1-3.

Kao, K.N. and M.R. Michayluk. 1974. A method for high frequency intergeneric fusion of plant protoplasts. Planta 115: 355-367.

Miller, W. et al. 2008. Sequencing the nuclear genome of the extinct woolly mammoth. Nature 456: 387-390.

Noonan, J.P. et al. 2006. Sequencing and analysis of Neanderthal genomic DNA. Science 314: 1113-1118.

Oleszczuk, J.J., Keith, D.M., and Keith, L.G. 1999. Projections of population-based twinning rates through the year 2100. Journal of Reproductive Medicine 44: 913-921.

Piña-Aguilar, R.E. et al. Revival of extinct species using nuclear transfer: Hope for the mammoth, true for the Pyrenean ibex, but is it time for "conservation cloning"? Cloning and Stem Cells 11: 341-346.

Vaughn, D., and Strathmann, R.R. Predators induce cloning in echinoderm larvae. Science 319: 1503.

Zorich, Z. 2010. Should we clone Neanderthals? Archaeology 63 (March/April).

We're All Mutants

Araten, D.J. et al. 2005. A quantitative measurement of the human somatic mutation rate. Cancer Research 65: 8111-8117.

Babic, A., Lindner, A.B., Vulic, M., Stewart, E.J., and Radman, M. 2008. Direct visualization of horizontal gene transfer. Science 319: 1533-1536.

Ellegren, H. 2002. Human mutation — blame (mostly) men. Nature 31: 9-10.

Goriely, A., McVean, G.A., Röjmyr, M., Ingemarsson, B., and Wilkie, A.O.M. 2003. Evidence for selective advantage of pathogenic FGFR2 mutations in the male germ line. Science 301: 643-646.

Griebel, C.P., Halvorsen, J., Golemon, T.B., and Day, A.A. 2005. Management of spontaneous abortion.

American Family Physician 72: 1243-1250.

Kaiser, J. 2006. First pass at cancer genome reveals complex landscape. Science 313: 1370.

Kaiser, J. 2008. A detailed genetic portrait of the deadliest human cancers. Science 321: 1280-1281.

Leitch, A.R., and Leitch, I.J. 2008. Genomic plasticity and the diversity of polyploidy plants. Science 320: 481-483.

Lynch, M. 2010. Rate, molecular spectrum, and consequences of human mutation. Proceedings of the National Academy of Sciences USA. 107: 961-968.

Moran, N.A., and Jarvik, T. 2010. Lateral transfer of genes from fungi underlies carotenoid production in aphids. Science 328: 624-629.

Kao, K.N. and Michayluk, M.R. 1974. A method for high frequency intergeneric fusion of plant protoplasts. Planta 115: 355-367.

Rosenberg, S.M. and Hastings, P.J. 2003. Modulating mutation rates in the wild. Science 300: 1382-1383.

Xue, Y., et al. 2009. Human Y chromosome base-substitution mutation rate measured by direct sequencing in a deep-rooting pedigree. Current Biology 19: 1453-1457.

Yamaguchi, Y. and Gojobori, T. 1997. Evolutionary mechanisms and population dynamics of the third variable envelope region of HIV within single hosts. Proceedings of the National Academy of Sciences USA. 94: 1264-1269.

Crossing the Species Boundary

Alper, H., Moxley, J., Nevoigt, E., Fink, G.R., and Stephanopoulos, G. 2006. Engineering yeast transcription machinery for improved ethanol tolerance and production. Science 314: 1565-1568.

American Society of Plant Biologists. ASPB statement on plant genetic engineering. www.aspb.org/publicaffairs/aspbgestatement.cfm.

Antunes, M.S., et al. 2011. Programmable ligand detection system in plants through a synthetic signal transduction pathway. PLoS ONE 6: 1-11.

Cohen, J. 2007. Building an HIV-proof immune system. Science 317: 612-614.

Enserink, M. 2008. Tough lessons from golden rice. Science 320: 468-471.

Finkel, E. 2008. Australia's new era for GM crops. Science 321: 1629.

Friedmann, T., Rabin, O., and Frankel, M.S. 2010. Gene doping and sport. Science 327: 647-648.

Greenpeace. 2010. Counting the costs of genetic engineering. www.greenpeace.org/international/en/publications/reports/counting-the-costs-of-genetic/.

Guo, J., and Xin, H. 2006. Splicing out the West. Science 314: 1232-1235.

Kaiser, J. 2008. Two teams report progress in reversing loss of sight. Science 320: 606-607.

Marvier, M., McCreedy, C., Regetz, J., and Kareiva, P. 2007. A meta-analysis of effects of Bt cotton and maize on nontarget invertebrates. Science 316: 1475-1477.

Morgan, R.A., et al. 2006. Cancer regression in patients after transfer of genetically engineered lymphocytes. Science 314: 126-129.

National Academies of Science. 2010. The impact of genetically engineered crops on farm sustainability in the United States — Report in Brief. www.nap.edu/catalog.php?record_id=12804.

Normile, D. 2008. Reinventing rice to feed the world. Science 321: 330-333.

Potrykus, I. 2010. Regulation must be revolutionized. Nature 466: 561.

Sierra Club. 2001. Genetic engineering: Genetic engineering at a historic crossroads. Sierra Club Genetic Engineering Committee Report. March.

Sinha, G., 2007. GM technology develops in the developing world. Science 315: 182-183.

Snow, A., Andow, D., Gepts, P., Hallerman, E., Power, A., Tiedje, J., and Wolfenbarger, L. 2005. Genetically engineered organisms and the environment: Current status and recommendations. Ecological Applications 15: 377–404.

Stokstad, E. 2008. GM papaya takes on ringspot virus and wins. Science 320: 472.

U.S. Department of Agriculture, National Agricultural Statistics Service (NASS). 2011. Acreage. http://usda.mannlib.cornell.edu/usda/nass/Acre//2010s/2011/Acre-06-30-2011.pdf#page=27. Viewed Sept. 24, 2011.

Wu, K., Lu, Y., Feng, H., Jiang, Y., and Zhao, J. Suppression of cotton bollworm in multiple crops in China in areas with Bt toxin – containing cotton. Science 321: 1676-1678.

Humans by Design

American Association for the Advancement of Science. 2006. Good, better, best: The human quest for enhancement. Summary report of an invitational workshop. Scientific Freedom, Responsibility and Law Program. www.aaas.org/spp/sfrl/projects/human_enhancement/#ST.

Andrews, L.B. 1999. The clone age. New York: Holt.

Centers for Disease Control and Prevention. 2006. National survey of family growth. http://www.cdc. gov/nchs/nsfg/abc_list_i.htm#infertilityservices.

Dickinson, E. 1858-1884. Selected poems. New York: Dover, 1990.

Huxley, J. 1957. New bottles for new wine. New York: Harper.

King, Jr., M.L. Dec. 11, 1964. Nobel lecture.

Merzenich, H., Zeeb, H., and Blettner, M. 2010. Decreasing sperm quality: A global problem? BMC Public Health 10: 24.

Nietzsche, F. 1892. Thus spoke Zarathustra. New York: Penguin, 1978.

Schrödinger, E. 1944. What is life? Cambridge: Cambridge University Press, 2002.

The Dawn of Virtual Beings

Alighieri, D. 1321/1961. The Purgatorio. Trans. John Ciardi. New York: Penguin.

Fagg, A.H. et al. 2007. Biomimetic brain machine interfaces for the control of movement. Journal of Neuroscience 27: 11842-11846.

Garreau, J. 2005. Radical evolution. New York: Broadway Books.

Lebedev, M.A., and Nicolelis, M.A.L. 2006. Brain-machine interfaces: Past, present and future. Trends in Neurosciences 29: 536-546.

Schrödinger, E. 1944. What is life? Cambridge: Cambridge University Press, 2002.

U.S. Department of Energy. 2009. Artificial retina project. http://artificialretina.energy.gov/.

Cosmic Evolution

Burroughs, W. 1980. The four horsemen of the

apocalypse. Lecture to the Eco-Technic Institute. http://digitalseance.wordpress.com/2007/10/30/william-s-burroughs-the-four-horsemen-of-the-apocalypse/.

Conway, A., Gilmour, I., Jones, B.W., Rothery, D.A., Sephton, M.A., and Zarnecki, J.C. 2004. An introduction to astrobiology. Cambridge: Cambridge University Press.

Dangerous Films. 2008. When we left Earth: The NASA missions. Produced in collaboration with NASA. Series producer, Kate Botting. Series writer, Ed Fields.

Kunitz, S. 2000. The collected poems. New York: Norton.

Kerr, R.A. 2010. Growing prospects for life on Mars divide astrobiologists. Science 330: 26-27.

NASA. 2003. Space life sciences research—1965-2003. NASA Ames Research Center.

NASA. 2007. Societal impact of spaceflight. Ed. S.J. Dick and R.D. Launius. Washington, D.C.: NASA.

National Geographic Society. 1969. First explorers on the moon. National Geographic 136: 735-797.

The Clearing Fog

Alighieri, D. 1321/1982. The Inferno. Trans. John Ciardi. New York: Penguin.

Huxley, J. 1942. Evolution: The modern synthesis. New York: Harper.

Palumbi, S. 2001. Humans as the world's greatest evolutionary force. Science 293: 1786-1790.

Schrödinger, E. 1944. What is life? Cambridge: Cambridge University Press, 2002.

Wilson, E.O. 2002. The future of life. New York: Vintage Books.

ABOUT THE AUTHOR

Dr. Jeffrey Scott Coker is Associate Professor of Biology and the Director of General Studies at Elon University in North Carolina. He has published widely on biology and education, appearing in The Futurist, American Biology Teacher, Biotechniques, Charlotte Observer, News and Observer, Writer's Journal, Physiologia Plantarum, and many other newspapers and academic journals. He has also led educational efforts for several state and national scientific societies. Reinventing Life originated as his popular science course at Elon University.

CPSIA information can be obtained at www.ICGtesting.com
Printed in the USA
LVOW10s0021240816

501571LV00034B/520/P